THE MAKING OF A NEUROMORPHIC VISUAL SYSTEM

THE MAKING OF A NEUROMORPHIC VISUAL SYSTEM

By

Christoph Rasche
Department of Psychology, Penn State University, USA

 Springer

Cover illustration: The chip layout was provided by Giacomo Indiveri and contains circuitry of integrate-and-fire neurons and synapses.

Library of Congress Cataloging-in-Publication Data

Rasche, Christoph, 1970-
   The making of a neuromorphic visual system / by Christoph Rasche.
      p. cm.
   Includes bibliographical references and index.
   ISBN 0-387-23468-3
   1. Neural networks (Computer science) 2. Visual pathways. 3. Integrated
circuits--Design and construction. I. Title.

   QA76.87.R37 2005
   006.3'2--dc22

                                                          2004058992

Printed in the United States of America.

9 8 7 6 5 4 3 2 1   SPIN 11323938

springeronline.com

Contents

1 Seeing: Blazing Processing Characteristics   **1**
1.1 An Infinite Reservoir of Information   1
1.2 Speed   1
1.3 Illusions   2
1.4 Recognition Evolvement   2
1.5 Basic-Level Categorization   3
1.6 Memory Capacity and Access   5
1.7 Summary   5

2 Category Representation and Recognition Evolvement   **7**
2.1 Structural Variability Independence   7
2.2 Viewpoint Independence   7
2.3 Representation and Evolvement   9
     2.3.1 Identification Systems   10
     2.3.2 Part-based Descriptions   11
     2.3.3 Template Matching   13
     2.3.4 Scene Recognition   13
2.4 Recapitulation   14
2.5 Refining the Primary Engineering Goal   15

3 Neuroscientific Inspiration   **17**
3.1 Hierarchy and Models   17
3.2 Criticism and Variants   20
3.3 Speed   23
3.4 Alternative 'Codes'   25
3.5 Alternative Shape Recognition   27
3.6 Insight from Cases of Visual Agnosia   29
3.7 Neuronal Level   31
3.8 Recapitulation and Conclusion   35

4 Neuromorphic Tools   **37**
4.1 The Transistor   37
4.2 A Synaptic Circuit   38
4.3 Dendritic Compartments   39
4.4 An Integrate-and-Fire Neuron   40
4.5 A Silicon Cortex   40
4.6 Fabrication Vagrancies require Simplest Models   42
4.7 Recapitulation   42

5 Insight From Line Drawings Studies   **45**
5.1 A Representation with Polygons   45
5.2 A Representation with Polygons and their Context   49
5.3 Recapitulation   51

6 Retina Circuits Signaling and Propagating Contours   **55**
   6.1 The Input: a Luminance Landscape 55
   6.2 Spatial Analysis in the Real Retina 55
      6.2.1 Method of Adjustable Thresholds 57
      6.2.2 Method of Latencies 58
   6.3 The Propagation Map 58
   6.4 Signaling Contours in Gray-Scale Images 60
      6.4.1 Method of Adjustable Thresholds 60
      6.4.2 Method of Latencies 60
      6.4.3 Discussion 64
   6.5 Recapitulation 64

7 The Symmetric-Axis Transform   **67**
   7.1 The Transform 67
   7.2 Architecture 68
   7.3 Performance 70
   7.4 SAT Variants 74
   7.5 Fast Waves 74
   7.6 Recapitulation 75

8 Motion Detection   **77**
   8.1 Models 77
      8.1.1 Computational 77
      8.1.2 Biophysical 77
   8.2 Speed Detecting Architectures 79
   8.3 Simulation 81
   8.4 Biophysical Plausibility 83
   8.5 Recapitulation 85

9 Neuromorphic Architectures: Pieces and Proposals   **87**
   9.1 Integration Perspectives 87
   9.2 Position and Size Invariance 89
   9.3 Architecture for a Template Approach 92
   9.4 Basic-Level Representations 94
   9.5 Recapitulation 95

10 Shape Recognition with Contour Propagation Fields   **97**
   10.1 The Idea of the Contour Propagation Field 97
   10.2 Architecture 98
   10.3 Testing 100
   10.4 Discussion 104
   10.5 Learning 107
   10.6 Recapitulation 109

11 Scene Recognition   111

11.1 Objects in Scenes, Scene Regularity 111
11.2 Representation, Evolvement, Gist 111
11.3 Scene Exploration 113
11.4 Engineering 115
11.5 Recapitulation 116

12 Summary   117

12.1 The Quest for Efficient Representation and Evolvement . 117
12.2 Contour Extraction and Grouping 121
12.3 Neuroscientific Inspiration 121
12.4 Neuromorphic Implementation 122
12.5 Future Approach 122

Terminology   125

References   129

Index   137

Keywords 137
Abbreviations 139

Preface

Arma virumque cano, Trojae qui primus ab oris
Italiam fato profugus, Laviniaque venit
litora.

This is the beginning of Ovid's story about Odysseus leaving Trojae to find his way home. I here tell about my own Odysee-like experiences that I have undergone when I attempted to simulate visual recognition. The Odysee started with a structural description attempt, then continued with region encoding with wave propagation and may possibly continue with a mixture of several shape description methods. Although my odyssey is still under its way I have made enough progress to convey the gist of my approach and to compare it to other vision systems.

My **driving intuition** is that visual category representations need to be loose in order to be able to cope with the visual structural variability existent within categories and that these *loose representations* are somehow expressed as neural activity in the nervous system. I regard such loose representations as the cause for experiencing visual illusions and the cause for many of those effects discovered in attentional experiments. During my effort to find such loose representations, I have made sometimes unexpected experiences that forced me to continuously rethink my approach and to abandon or turn over some of my initially strongly believed viewpoints. The book therefore represents somewhat the odyssey through different attempts: At the beginning I pursued a typical structural description scheme (chapter 5), which eventually has turned into a search of a mixture of shape description methods using *wave-propagating* networks (chapter 10). What the exact nature of these representations should look like, is yet still unclear to me, but one would simply work towards it by constructing, testing and refining different architectures. I regard the *construction* of a visual system therefore as a *stepwise* process, very similar to the invention and evolutionary-like refinement of other technical systems like the automobile, airplane, rocket or computer. In order to build a visual system that processes with the same or similar efficiency, I believe that it is worth to understand how the human visual system may achieve this performance on a behavioral, on an architectural as well as on a network level. To emulate the envisioned mechanisms and processes with the same swiftness, it may be necessary to employ a substrate that can cope with the intensity of the demanded computations, for example the here mentioned neuromorphic analog circuits (chapter 4).

More specifically, I have approached the **design endeavor** by firstly looking at some behavioral aspects of the seeing process. Chapter 1 lists these observations, which help to identify the *motor* of vision, the basic-level categorization process, and which help to define its very basic operation. I consider the understanding and construction of this categorization process as a starting point to engineer a visual system. Chapter 2 describes two more characteristics of the basic-level categorization process, with which I review some of the past and current vision systems. Chapter 3 reviews the progress made so far in the neuroscientific search for the biological architecture. Chapter 4 mentions the necessary neuromorphic analog circuits for the processes I simulate. Chapter 5 reports about a computer vision simulation study using line drawing objects, from which I gained the insight that *region* (or space) is important information for representation and evolvement. I then turn towards gray-scale images. The idea of region encoding is translated into the neuromorphic language, whereas chapter 6 presents retinal circuits that signal contours in gray-scale images, and chapter 7 introduces the networks that perform Blum's *symmetric-axis transform*. With the obtained symmetric-axes one could already carry out a substantial amount of categorization using a computer vision back-end that associates the obtained axes - it would be a *hybrid categorization system*. Chapter 8 makes a small detour into motion detection, specifically speed detection. Chapter 9 is a collection of neuromorphic architectures and thoughts on structural description, template matching, position and size invariance, all of which is relevant when one tries to build a fully neuromorphic visual system. An instantiation of those ideas is presented in chapter 10, which describes a novel region encoding mechanism, and which has the potential to be the fundament for an efficient shape description. The experiences made thus far, are translated to the issue of scene recognition, which is summarized in chapter 11. The final chapter, number 12, recapitulates my journey and experiences.

The **inspiring literature** for my vision approach was Palmer's book (1999), which I consider as indispensable reading for anyone who tries to understand representational issues in vision from an interdisciplinary viewpoint. Some of the points I make in this discourse are much broader embedded in Palmer's book. The inspiring literature for my 'neuromorphic' realization was Blum's thoughts on the possibility of the brain working as a broadcast-receiver principle (1967), an idea that has never been seriously explored, but which I pick up here, because it solves certain problems elegantly.

A word on **terminology**: As Fu already noted (Lee and Fu, 1983), visual recognition and representation is difficult in problem formulation and in computational methodology. I have therefore created a short terminology section (page 119), that hopefully clarifies some of the terms which are floating throughout the chapters and other vision literature, and that puts those terms into perspective.

Acknowledgments

The breadth of this work would not have been possible without the necessary broad education and support that I have received from my previous advisors. I am deeply indepted to thank:

Rüdiger Wehner (Institute of Zoology, University of Zürich, neuromorphic engineer of desert ants), for directing me towards computational neurobiology.

Rodney Douglas (Institute of Neuroinformatics, ETH Zürich, neuromorphic engineer of the neocortex), for putting me behind the silicon neuron project.

Christof Koch (Caltech, Pasadena, neuromorphic engineering consciousness one day), for crucial support when I was setting sails for my visual odyssey.

I would like to particulary thank Michael Wenger (Penn-State, Pennsylvania), with whose support the writing of this book went much faster than expected. I greatly appreciated the feedback on some of my early manuscripts by Richard Hahnloser (ETH Zürich). Part of my writing and reasoning skills I owe to Peter König (now at Universität Osnabrück). I also enjoyed support by Miguel Eckstein (UC Santa Barbara).

Christoph Rasche
Penn-State (University Park), Summer 2004

1 Seeing: Blazing Processing Characteristics

We start by listing a few, selected behavioral phenomena of the vision process, which help us to define its very basic operation.

1.1 An Infinite Reservoir of Information

When we look at a visual scene, like a room or outdoor scene, we can endlessly explore its content using eye movements. During the course of this exploration, we find an infinite number of details like different colors, textures, shapes of objects and object parts and their structural relations. The saying 'A picture is more worth than a 1000 words' is an understatement of the enormous information content in a scene. This endless amount of information is scientifically well pointed out by Yarbus' studies on human eye movements (Yarbus, 1967). Yarbus has traced the fixation points of a person when he/she was browsing the photo of a room scene containing people engaged in a social situation. Yarbus recorded this sequence of eye movements for a few minutes, thereby giving the subject a different task for each recording. In an unbiased condition, the observer was instructed to investigate the scene in general. In other conditions, the observer was given for example the task to judge the ages of the people present in the scene. Each condition resulted in a very *distinct fixation pattern* in which fixation points are often clustered around specific features. Hence, the information content of a scene is an infinite reservoir of interesting details, whose thorough investigation requires an extensive visual search.

1.2 Speed

Probably one of the most amazing characteristics of visual processing is its operation speed. When we look at a picture, we instantaneously comprehend its rough content. This property is exploited for example by makers of TV commercials, who create fast-paced TV commercials in order to minimize broadcast costs. Potter has determined the speed with which humans are able to apprehend the gist of a scene or object using the rapid-serial-visual-presentation technique (Potter, 1976). Before an experiment, a subject was shown a target picture. The subject was then presented a rapid sequence of different images, of which one could be the target picture. At the end of a sequence, the subject had to tell whether the sequence contained the target picture or not. When the presentation rate was four pictures a second (every 250ms), subjects had little problems to detect the target picture. For shorter intervals, the recognition percentage would drop, but still be significantly above chance level even for presentation intervals of

100ms only. This time span is way less than the average fixation period between eye movements which is around 200 to 300ms.

1.3 Illusions

Given the speed of recognition, one may think we sometimes err in our interpretation of a scene or object? Indeed, it happens frequently: we often mistake an object for another one, for example either because it is out of focus or because we are in a rush or because it is an unusual view. But most of the time we are not particularly aware of these minor mistakes, because they are immediately corrected by the continuous stream of visual analysis. Certain visual illusions expose this property very distinctively. When we see an illusion, like the *Impossible Trident* (figure 1), we immediately have an idea of what the structure is about. After a short while of inspection though, we realize that the structure is impossible. Escher's paintings - possessing similar types of illusions - are an elegant example of how the visual system can be *tricked*. One may therefore regard the visual system as faulty or as processing to hastily. Yet, it is more likely that it was built for speed, a property which is of greater importance for survival than a slow and detailed reconstruction.

Figure 1: Impossible Trident. Illusions like this one are able to explicitly trick the recognition process. They evidence that representations are structurally loose.

1.4 Recognition Evolvement

Based on the above three mentioned properties, one may already start to characterize the recognition process. Despite the enormous amount of information in a scene, the visual system is able to understand its rough content almost instantaneously. Thus, there must be a process at work, that is able to organize the information suitable for quick understanding. Given that this process can be deceited, one may infer that it is structurally not accurate in its *evolvement* or in the type of representations it uses - an inaccuracy that is exposed only rarely and that can quickly be corrected by swift, subsequent analysis. Although we believe that this recognition evolvement is a fluent process, it makes sense to divide it into *two separate stages* and to label it with commonly used terms for reason of clarity, see figure 2. In a perceptual stage, visual structure is initially guessed by using some inaccurate representation. This rapid association in turn triggers a cognitive stage employing a semantic representation, that allows to confirm or verify the perceived structure. Based on similar reflections about visual illusions, Gregory has proposed are more refined concept of the recognition process (Gregory, 1997), but our present, simpler proposal suffices for the beginning.

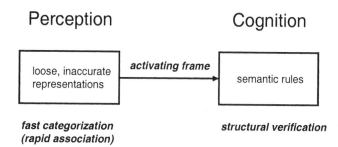

Figure 2: A simplified, discretized illustration of the fluent object recognition evolvement process. In a 'perceptual' stage, the system quickly categorizes the object using loose representations, which triggers a frame. In a 'cognitive' stage, semantic rules verify the perceived structure.

The idea of a continuous recognition evolvement fits well with the idea of *frames*. Frames are collections of representations, which are retrieved when we have recognized the gist of a scene or object for example. Frames would allow us to browse a scene much quicker than if they were not existent. The idea has been put forward by different researchers from fields like Psychology and Artificial Intelligence. The most specific and concise formulation was given by Minsky (Minsky, 1975) (and see references therein). We relate the idea of frames to our envisioned scheme as follows: the perceptual stage (or *perceptual category representations*) would trigger such a frame containing a set of semantic rules describing the representations of objects or scenes in a structurally exhaustive manner. A more general term of this type of guidance would be 'top-down' influence.

1.5 Basic-Level Categorization

The process that enables to quickly organize visual structure into useful information packages is termed the *basic-level categorization* process (Rosch et al., 1976). Rosch et al. carried out experiments, in which humans had to name objects that they were presented. The experiments showed that humans classify objects into categories like car, table and chair, which Rosch et al. termed basic-level categories. They found other levels of categories as well (figure 3). On a more abstract level, there are categories like tool, vehicle or food, which they termed super-ordinate categories. On a more specific level, there are categories like sports car, kitchen table or beach chair, which they termed subordinate categories. In the hierarchy shown in figure 3 we have added another level, the identity level, at which one recognizes objects that represent particular instances of categories, e.g. a car model or a chair model. If one looks at different instances of the same category, then one realizes there are many, slight structural differences between them. For example a desktop can have one or two chests of drawers, the chest can have a different number of drawers and so on. The representation of visual objects must therefore be something *loose* in order to be able to deal with such variability. This loose representation may be the reason why the recognition system is prune to structural visual illusion. But that may not even be the proper formulation of this characteristic: it may very well be that representations have to be inaccurate and loose in order to be able to efficiently categorize. In some sense, the *'structural inaccuracy' may be a crucial strength.*

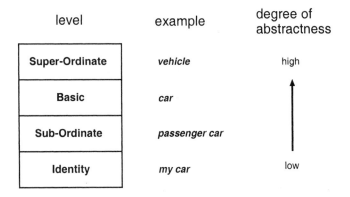

level	example	degree of abstractness
Super-Ordinate	*vehicle*	high
Basic	*car*	
Sub-Ordinate	*passenger car*	
Identity	*my car*	low

Figure 3: Category levels in the visual system.

When we perform a categorization, the recognition process has likely ignored a lot of details of that object. The object has been perceived with some sort of abstract representation, which we believe is the cause for experiencing visual illusions and which is the cause for the many effects seen in attentional experiments, like the lack of full understanding of the image (O'Regan, 1992; Rensink, 2000). This abstract representation is what guides our seeing process.

Objects of the same basic-level category can come in different textures, colors and parts. From this variety of visual cues, it is generally shape that retains most similarity across the instances of a basic-level category and that is the cue we primarily focus on in this book.

1.6   Memory Capacity and Access

Another stunning characteristic of the visual system is its memory capacity. We swiftly memorize most new locations where we have been to, we instantaneously memorize a torrent of image sequences of a movie or TV commercial. And we can easily recall many of these images and sequences even after a long period of time. Standing et al. have shown these immense storage capacities and stunningly fast access capabilities by presenting subjects with several hundreds of pictures, most of which could be recalled next day or later (Standing et al., 1970).

There seems to be a paradox now. On the one hand, when we see a novel image, we comprehend only a fraction of its information content and it would require a visual search to accurately describe a scene. On the other hand, we are able to memorize a seemingly infinite number of images relatively swiftly. Ergo, if we see only a fraction of the image, then it should be surprising that we are still able to distinguish it so well from other images. The likeliest explanation is that with a few glances at an image, one has swallowed enough information, that makes the percept distinct from most other scenes. Speaking metaphorically, a single scoop from this infinite information reservoir apparently suffices to make the accumulated percept distinguishable from many other pictures.

1.7   Summary

The visual machinery organizes visual structure into classes, so called basic-level categories. It does this fast and efficiently, but structurally inaccurate as evidenced by visual illusions. The type of representation it uses may be inaccurate and loose, in order to be able to recognize novel objects of the same category that are structurally somewhat different. Because of this representational inaccuracy, the visual system occasionally errs, but that is often quickly overplayed by rapid continuous analysis. The machinery ignores many structural details during the categorization process. Still it retains sufficient information to be distinct from other images.

We understand this as the coarsest formulation of the seeing process and it suffices already to envisage how to construct a visual system. We believe that the *primary engineering goal* should be to firstly build this categorization process. In a first construction step, one would solely focus on the perceptual stage (left side in figure 2): this stage would categorize objects using only some sort of inaccurate, perceptual representation. In a second step, one may think about how to represent semantic knowledge, that would allow for verification of the perceived structure (right side in figure 2). The first step is already challenging enough and that is what this book aims at: working towards a neuromorphic architecture that carries out the perceptual stage performing swift categorization. In the next chapter we are trying to specify the nature of this perceptual stage by looking closer at some aspects of the basic-level categorization process.

2 Category Representation and Recognition Evolvement

We here list two more aspects of the recognition process, the aspect of structural variability independence and the aspect of viewpoint independence (Palmer, 1999). With these two aspects in mind, we characterize previous and current vision systems and it will allow us to better outline the systematics of our approach.

2.1 Structural Variability Independence

We have already touched the aspect of structural variability independence in the previous chapter. Here we take a refined look at it. Figure 4 shows different instances of the category 'chair', with the goal to point out the structural variability existent within a category. We intuitively classify the variability into three types:

a) *Part-shape variability*: the different parts of a chair - leg, seat and back-rest - can be of varying geometry. The legs' shape for example can be cylindrical, conic or cuboid, sometimes they are even slightly bent. The seating shape can be round or square like or of any other shape, so can the back-rest (compare chairs in figure 4a).

b) *Part-alignment variability*: the exact alignment of parts can differ: the legs can be askew, as well as the back-rest for more relaxed sitting (top chair in figure 4b). The legs can be exactly aligned with the corners of the seating area, or they can meet underneath it. Similar, the back-rest can align with the seating area exactly or it can align within the seating width (bottom chair in figure 4b).

c) *Part redundancy*: there are sometimes parts that are not necessary for categorization, as for example the arrest or the stability support for the legs (figure 4c). Omitting these parts does still lead to proper categorization.

Despite this variability, the visual system is able to categorize these instances: the process operates independent of structural variability. A chair representation in the visual system may therefore not depend on exact part shapes or exact alignments of parts. It may neither contain any structures that are not absolutely necessary for categorization. The corresponding category representation would therefore be something very loose and flexible. The degree of looseness would depend on the degree of variability found in a category. For example, the category chair certainly requires a larger degree of looseness than the category book or ball.

2.2 Viewpoint Independence

Another aspect of recognition is its viewpoint independence. We are able to recognize an object from different viewpoints despite the dif-

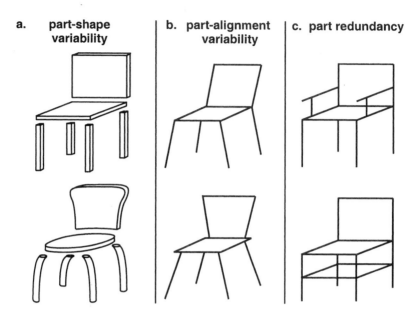

Figure 4: Intuitive classification of structural variability in the category chair. a. Part-shape variability. b. Part-alignment variability. c. Part redundancy. The category representation must be something loose and flexible.

ferent 2D appearance of the object's structure for any given viewpoint. The viewpoints of an object can be roughly divided into *canonical* and *non-canonical* (Palmer et al., 1981). Canonical viewpoints exhibit the object's typical parts and its relations, like the chairs seen in the left of figure 5. In contrast, non-canonical viewpoints exhibit only a fraction of the object's typical parts or show the object in unexpected poses, and are less familiar to the human observer, like the chairs seen in the right of figure 5.

In our daily lives we see many objects primarily from canonical viewpoints, because the objects happen to be in certain poses: chairs are generally seen on their legs or cars are generally on their wheels. Canonical viewpoints can certainly be recognized within a single glance (Potter, 1975; Thorpe et al., 1996; Schendan et al., 1998). In contrast, non-canonical viewpoints are rare and one can assume that the recognition of non-canonical viewpoints requires more processing time than a single glance. Recognizing a non-canonical viewpoint may consist of a short visual search using a few saccades, during which textural details are explored; or the perceived structure of the object is transformed in some way to find the appropriate category (Farah, 2000). Behavioral evidence from a person with visual agnosia suggests that non-canonical views are indeed something unusual (Humphreys and Riddoch, 1987a). The person is able to recognize objects in daily live without problems, yet struggles to comprehend non-canonical views of objects given in a picture. This type of visual disorder was termed perceptual categorization deficit, but pertains to the categorization of unusual (non-canonical) views only. One may conclude from this case, as Farah does, that such views do not represent any real-world visual tasks.

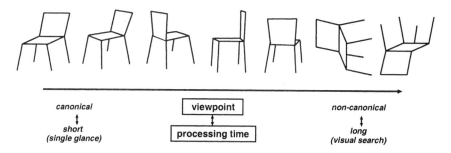

Figure 5: Different poses and hence viewpoints of a chair. Viewpoint and processing time are correlated. Left: canonical views that are quickly recognizable. Right: non-canonical views that take longer to comprehend, possibly including a saccadic visual search.

2.3   Representation and Evolvement

We now approach the heart of the matter: how are we supposed to represent categories? Ideally, the design of a visual system starts by defining the nature of *representation* of the object or category, for example the object is described by a set of 3D coordinates or a list of 2D features. This representation is sometimes also called the object model. In a second step, after defining the representation, a suitable reconstruction method is contrived that extracts crucial information from the image, which in turn enables the corresponding category. One may call this object *reconstruction* or *evolvement*. Such approaches were primarily developed from the 60's to the 80's, but are generally not extendable into real-world objects and gray-scale images. Recent approaches have taken a heuristic approach, in which the exact representation and evolvement is found by testing.

Most of these systems - whether fully designed or heuristically developed - start with some sort of contour extraction as the first step, followed by classifying contours and relating them to each other in some way to form higher features, followed by finding the appropriate category. We here review mainly Artificial Intelligence (computer vision) approaches and some psychological approaches . Neural network approaches are mentioned in the next chapter.

2.3.1 Identification Systems

Early object recognition systems aimed at identifying simple building blocks from different viewpoints. Because they intended to do that precisely, the object model was defined as a set of corner points specified in a 3D coordinate system. Robert devised such a system performing this task in roughly three steps (figure 6,(Robert, 1965)): Firstly, contours were extracted and 2D features formed. Secondly, these extracted 2D features were matched against a set of stored 2D features that would point towards a specific object. Finally, each of those object models, whose 2D features were successfully matched in the second step, were matched against the contours, determining so the object identity. With the identified object model it is possible to find the object's exact pose in 3D space.

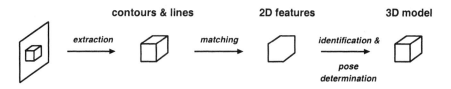

Figure 6: Roberts identification and pose determination system. The object was represented as a set of 3D coordinates representing the corners of a building block. Recognition evolved firstly by extracting contours and lines, followed by a matching process with stored 2D features, followed by eventual matching some of the possible models against the image.

Many later systems have applied this identification and pose determination task to more complex objects using variants, elaborations and refinements of Roberts' scheme (Brooks, 1981; Lowe, 1987; ULL-MAN, 1990; Grimson, 1990). Some of them are constructed to serve as vision systems for part assembly in industry performed by roboters. Some of them are able to deal with highly cluttered scenes, in which the object identity is literally hidden in a mixture of lines. These systems do so well with this task, they may even surpass the performance of an untrained human 'eye'.

All of these systems work on the identity level (figure 3, chapter 1). They do not categorize and therefore do not deal with structural variability and have in some sense 'only' dealt with the viewpoint independence aspect. They have been applied in a well defined environment with a limited number of objects. The real world however contains an almost infinite number of different objects, which can be categorized into different levels. The structural variability that one then faces therefore demands different object representations and possibly a different recognition evolvement.

The construction of such pose-determining systems may have also influenced some psychological research on object recognition, which attempts to uncover that humans recognize objects from different viewpoints by performing a similar transformational process as these computer vision systems do (e.g. see (Tarr and Bulthoff, 1998; Edelman, 1999) for a review).

2.3.2   Part-based Descriptions

Part-based approaches attempt to describe objects by a set of forms or 'parts', arranged in a certain configuration: it is also called a structural description approach (figure 7).

Guzman suggested a description by 2D features (Guzman, 1971). In his examples, an object is described by individual shapes: For example, a human body is described by a shape for the hand, a shape for the leg, a shape for the foot and so on. These shapes were specified only in two dimensions. Figure 7 shows a leg made of a shape for the leg itself and a shape for a shoe. Guzman did not specifically discuss the aspect of structural variability independence, but considered that objects can have deformations like bumps or distortions and that despite such deformations the visual system is still able to recognize the object correctly. In order to be able to cope with such deformations, he proposed that representations must be sort of 'sloppy'. This aspect of 'deformation independence' is actually not so different from the aspect of structural variability independence.

Binford came up with a system that measures the depth of a scene by means of a laser-scanning device (Binford, 1971). His objects were primarily expressed as a single 3D volume termed 'generalized cones', which were individual to the object. For example the body of a snake is described as one long cone (Agin and BINFORD, 1976). Reconstruction would occur by firstly extracting contours, followed by determining the axis of the cones using a series of closely spaced cone intersections. The example in figure 7 shows a snake, which is represented by a single, winding cone. Binford did not specifically address the structural variability aspect.

Binford's system likely influenced Marr's approach to represent animal bodies by cylinders (Marr and Nishihara, 1978). The human body for example would be represented as shown in figure 7. Similar to Binford, Marr planned to reconstruct the cylinders by finding their axes: firstly, surfaces of objects are reconstructed using multiple cues like edges, luminance, stereopsis, texture gradients and motion, yielding the 2.5D 'primal sketch' (Marr, 1982); secondly, the axis would be reconstructed and put together to form the objects. Marr did not specifically address the aspect of structural variability either, but cylinders as part representations would indeed swallow a substantial amount of structural variability. The idea to reconstruct surfaces as a first step in recognition was emphasized by Gibson (e.g. (Gibson, 1950)).

Pentland described natural objects like trees with superquadrics like diamonds and pyramidal shapes (Pentland, 1986) (not shown in figure 7).

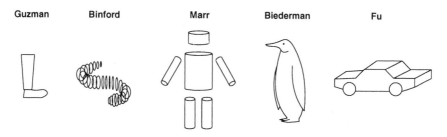

Figure 7: Object representations by parts. Guzman: individual 2D shapes. Binford: 'generalized cones'. Marr: cylinders. Biederman: geons. Fu: surfaces. Loosely redrawn from corresponding references given in text.

Fueled by the idea of a representation by 3D volumes, Biederman proposed an even larger set of 'parts' for representation, like cylinders, cuboids and wedges, 36 in total, which he called 'geons' (Biederman, 1987). The example in figure 7 shows a penguin made of 9 different such geons. To account for the structural variability, Biederman suggested that a category representation may contain interchangeable geons for certain parts. This may however run into a combinatorial explosion for certain categories, especially the ones with a high structural variability. The evolvement of the geons and objects would start with finding firstly vertex features.

These part-based approaches have never really been successfully applied to a large body of gray-scale images. One reason is, that it is computationally very expensive to extract the volumetric information of each single object part. Another reason is that the contour information is often fragmentary in gray-scale images and that this incomplete contour information does not give enough hints about the shape of 3D parts, although Marr tried hard to obtain a complete contour image (Marr, 1982). Instead of this 3D reconstruction, it is cheaper and easier to interpret merely 2D contours, as Guzman proposed it. Fu has done that using a car as an example (figure 7): the parallelograms that a car projects onto a 2D plane, can be interpreted as a surface (Lee and Fu, 1983). Still, an extension to other objects could not be worked out.

Furthermore, in most of these part-based approaches, the representations are somewhat chosen according to human interpretation of objects, meaning a part of the recognition system corresponds to a part in a real object, in particular in Guzman's, Marr's and Biederman's approach. But these types of parts may be rather a component of the semantic representation of objects (figure 2, right side). As we pointed out already, the perceptual representations we look for, do not need to be that elaborate (figure 2, left side). Nor do they need to rely on parts.

2.3.3   Template Matching

In a template matching approach, objects are stored as a 2D template and directly matched against the (2D) visual image. These approaches are primarily developed for detection of objects in gray-scale images, e.g. finding a face in a social scene or detecting a car in a street scene. Early attempts tried to carry out such detection tasks employing only a 2D luminance distribution, which was highly characteristic to the category. To find the object's location, the template is slid across the entire image. To match the template to the size of the object in the image, the template is scaled. Because this sliding and scaling is a computationally intensive search procedure, the developers of such systems spend considerable effort in finding clever search strategies.

Recent attempts are getting more sophisticated in their representations (Amit, 2002; Burl et al., 2001). Instead of using only the luminance distribution per se, the distribution is nowadays tendentially characterized by determining its local gradients, the differential of neighboring values. This gradient profile enables a more flexible matching. Such a vision system would thus run first a gradient detection algorithm and the resulting scene gradient-profile (or landscape) is then searched by the object templates. In addition, an object is often represented as a set of sub-templates representing significant 'parts' of objects. For instance, a face is represented by templates for the eyes and a template for the mouth. In some sense, these approaches move toward more flexible representations in order to be able to cope with the structural variability existent in categories. These systems can also perform very well, when the image resolution is low. In comparison, in such low resolution cases, a human would probably recognize the object rather with help of contextual information, that means that neighboring objects facilitate the detection of the searched object. Such contextual guidance can take place with frames.

2.3.4   Scene Recognition

The first scene recognition systems dealt with the analysis of building blocks like cuboids and wedges depicted in line drawings, so-called polyhedral scenes. Guzman developed a program that was able to segment a polyhedral scene into its building blocks (Guzman, 1969).

His study trailed a host of other studies refining and discussing this type of scene analysis (Clowes, 1971; Huffman, 1971; Waltz, 1975). The goal of such studies was to determine a general set of algorithms and rules that would effectively analyze a scene. However the explored algorithms and representations are difficult to apply to scenes and objects in the real world because their structure is much more variable.

Modern scene recognition attempts aim at the analysis of street scenes depicted in gray-scale images. A number of groups tries to form representations for objects made of simple features like lines and curves, and of a large set of rules connecting them (e.g. (Draper et al., 1996)). Evolvement would occur by a set of control feedback loops, searching for the correct match. These groups have faced the structural variability aspect and addressed it as follows: when they are confronted with a large variability, they 'sub-categorize' a basic-level category, moving thus toward an increasing number of 'templates'.

Many of these systems intend to recognize objects from gray-scale images that have a relatively low resolution. In these images, objects can appear very blurred and it is very difficult and probably even impossible to perform proper recognition without taking context into account, as the developers realized. The human observer has of course no problem categorizing such images, thanks to the power of frames that can provide rich contextual information. We have more on the subject of scene recognition in chapter 11.

2.4 Recapitulation

We summarize the different approaches with regard to their type of representations - whether they are specified in 2D or 3D - and their method of reconstruction (figure 8).

Some artificial intelligence approaches focused on object representations specified in a 3D dimensional coordinate system and they attempted to reconstruct the constituent 3D parts directly from the image, like Binford's and Marr's approach, as well as Brook's identification system (figure 8a). Roberts' and Lowe's system also represent objects in 3D, but evolvement was more direct by going via 2D features (figure 8b). Scene recognition approaches search for representations using merely simple 2D features and extensive feedback loops for matching (figure 8c). The most direct recognition systems are the template matching systems, which can be roughly labeled as 2D-2D systems (figure 8d). We also assign neural networks (NN) to that category, because many of them aim at a feature matching in some sense (chapter 3, section 3.1). The single arrow should indicate that evolvement is either direct (in case of templates) or continuous (for neural networks). Figure 8e refers to spatial transformations which we will also discuss in chapter 3.

In case of the identification systems, the representation and evolve-

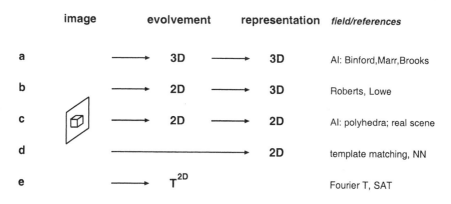

Figure 8: Summary of recognition systems, roughly ordered by evolvement strategies and representation type. Top (a): pure 3D approaches, the model as well as reconstruction occurred via 3D volumes. Bottom (f): representation and evolvement involving spatial transformations.

ment was defined beforehand. This worked well because the range of objects was finite and their environment was often well defined. The part-based approach also defined representation and evolvement ahead, but this has not led to general applicable systems. Their type of representations seemed to correspond to a human interpretation of objects and may therefore serve better as a cognitive representation (right side of figure 2). Because successful representations are difficult to define, approaches like template matching and scene recognition employ an exploratory approach.

2.5 Refining the Primary Engineering Goal

Given the large amount of variability, it is difficult to envision a category representation made of a fixed set of rigid features. Our proposal is to view a category representation as a loose structure: the shape of features as well as their relations amongst each other is to be formulated as a *loose skeleton*. The idea of loose representations has already been suggested by others. 1) Ullman has used fragmented template representations to detect objects depicted in photos (Ullman and Sali, 2000). 2) Guzman has also proposed that a representation needs to be loose (Guzman, 1971). He developed this intuition - as mentioned before - by reflecting on how to recognize an object, that has deformations like bumps or distorted parts. He termed the required representation as 'sloppy'. 3) Results from memory research on geographical maps suggests that human (visual) object representations are indeed fragments: Maps seem to be remembered as a collage of different spatial descriptors (Bryant and Tversky, 1999). Geographical maps are instances of the identity level (figure 3): Hence, if even an instance of an identity is represented as a loose collage, then one can assume that basic-level category representations are loose as well, if not even much looser. Loose representations can also provide a certain degree of viewpoint invariance. Because the structural relations are not exactly specified, this looseness that would enable to recognize objects from slightly different viewpoints. We imagine that this looseness is restricted to canonical views only. Non-canonical views likely trigger an alternate recognition evolvement, for instance starting with textural cues.

At this point we are not able to further specify the nature of representations, nor the nature of recognition evolvement. We will do this in our simulation chapters (chapters 5, 7 and 8). Because it is difficult to define a more specific representation and evolvement beforehand, our approach is therefore exploratory like the template and scene recognition systems, but with the primary focus on the basic-level categorization process. The specific goal is to achieve categorization of canonical views. Non-canonical views are not of interest because they are rare (section 2.2). Thus, the effort has to go into finding the neuromorphic networks that are able to deal with the structural variability. Furthermore, this system should firstly be explored using objects which are depicted at a reasonable resolution. Once this 'motor' of vision, the categorization process, has been established, then one would refine it and make it work on low-resolution gray-scale images or extend it to recognition of objects in scenes.

3 Neuroscientific Inspiration

Ideally one would understand how the real, biological visual system processes visual information and then one would mimic these mechanisms using the same networks. To gain such neuroscientific inspiration, we start by looking at the prevailing neuroscientific paradigm of visual processing, followed by reviewing some of the criticism it has drawn to. The criticism comes from the neuroscientific discipline itself, but also from psychological as well as from computational viewpoints.

Many of the experiments, giving us insight about the biological visual system, are carried out in monkeys (and sometimes even cats), but it is assumed that the human visual system has a functionally similar architecture allowing so for an analogy.

3.1 Hierarchy and Models

Neurophysiology The neuroscientific view of recognition can be termed a local-to-global evolvement, that is, an evolvement starting with small features and then gradually integrating toward global features and the eventual percept, see figure 9 (Palmer, 1999; Farah, 2000).

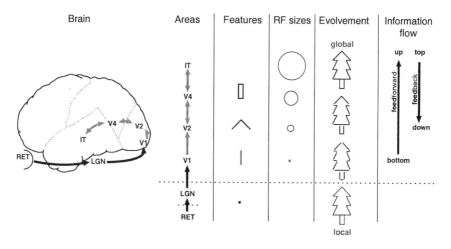

Figure 9: Schematic illustration of the (supposed) local-to-global recognition evolvement along the cortical hierarchy. Brain: Outline of the primate brain with important visual areas. Areas: The areas depicted as hierarchy. Features: Increasing complexity along the hierarchy. RF size: receptive field sizes. Evolvement: local-to-global. Supposed information flow: supposed flow in cortical areas.

In the retina, a visual image is analyzed point by point. Retinal ganglion cells respond to a small, circular spot of the visual field, the so-called receptive field (RF), by generating a firing rate that corresponds to the luminance value impinging its receptive field (Barlow, 1953; Kuffler, 1953). The thalamus seems to relay this point-wise analysis. In the primary visual cortex (V1), there exist orientation-selective cells that respond with a high firing frequency for a short piece of contour of a certain angle, or also called *orientation*. Their receptive field is elongated and they systematically cover the entire visual field, forming an organized structure that has been termed *orientation columns* (Hubel and Wiesel, 1962; Hubel and Wiesel, 1968), see figure 10c. Some of these orientation-selective cells respond to static orientation. Yet, most of them are motion-sensitive and respond to an oriented bar or edge moving across their RF, see figure 10a and b for a schematic summary. Higher cortical areas, like V2 and V4, have many cells responding to similar features like the ones in V1, but have a larger receptive field, thus covering a larger area of the visual field. In addition, and more interestingly, some of the cells in V2 and V4 respond also to more complex features, like angles, stars, concentric circles, radial features, polar plots and so on (e.g. (Hegde and Essen, 2000; Gallant et al., 1996; Pasupathy and Connor, 1999). Some of these features are schematically shown in figure 11. Cells in the inferior temporal cortex (IT) also respond to simple stimuli like oriented lines but some of them signal for even more complex shapes than the cells in areas V2 and V4 do (Gross et al., 1972; Tanaka et al., 1993), see figure 11 for some examples. Some of the IT cells signal for photographed objects (Kreiman et al., 2000). IT cells have large receptive field sizes and show some invariance to the exact position and size of the object or shape they respond to. Visual recognition may even continue into the prefrontal cortex, where cells are apparently involved in visual categorization (e.g. (Freedman et al., 2002), not shown in figure). In all these recordings, those cells were designated as 'feature detectors', that responded with a high firing rate, because it is believed that a '*rate code*' is the type of signal with which the neurons communicate with each other.

There is a number of reasons that led to this idea of a hierarchical '*feature integration*': One is, that the complexity of the detected features seemingly increases along the hierarchy. Another reason is, that these areas seem to be serially connected. A third reason is, that receptive field sizes are increasing from lower areas (e.g. V1) to higher areas (e.g. IT). This feature-integration scheme has been most explicitly formulated by Barlow, whereby he envisioned that the neuron is the fundamental perceptual unit responding to individual aspects of the visual field (Barlow, 1972). This 'vision' was given the term grandmother-cell theory, because Barlow's example object was a line-drawing picture of a grandmother.

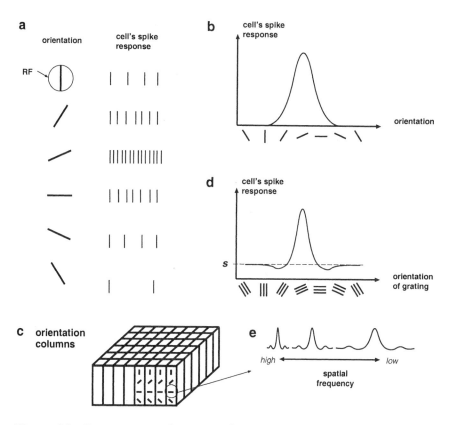

Figure 10: Orientation selectivity of V1 cells. a. Spiking of a V1 cell in response to different orientations. This cell prefers orientations of 66 degrees approximately, showing a high firing frequency for that orientation. b. Orientation-tuning curve for the cell in a. c. Orientation columns. d. V1 cell stimulated with oriented gratings. S: spontaneous firing rate. e. Additional dimension: resolutional scale (or spatial frequency).

Models Models, that mimic this hierarchy, have been contrived since the 60's, see (Rolls and Deco, 2002) for a history. Fukushima was the first to thoroughly simulate a hierarchical system applying it to recognition of digits (e.g. (Fukushima, 1988)). Recent models refine this concept (Edelman, 1999; Riesenhuber and Poggio, 1999). These models generally operate in a feed-forward (bottom-up) manner. Other researchers like Grossberg as well as Rolls and Deco have invented more elaborate models that are not necessarily as strictly hierarchical as sketched here. But all the models share the idea of feature integration (Francis et al., 1994; Bradski and Grossberg, 1995; Rolls and

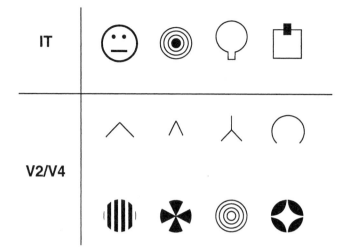

Figure 11: Feature selectivity of some of the cells in cortical areas V2, V4, IT. Most cells respond to simpler stimuli like in V1 but there are also cells that respond to complex stimuli like the ones displayed here. Features redrawn from corresponding references, see text.

Deco, 2002). And in all these models, the input-output function of a neuron expresses the idea of a rate code.

3.2   Criticism and Variants

Distributed Hierarchy and Representation   The idea of feature integration comes in variants. The one discussed in the previous subsection can be termed an evolvement by convergence of neuronal connections ((Rolls and Deco, 2002), figure 12a, 'single neuron'). One aspect that can be criticized on this scheme is the strictly hierarchical interpretation of the cortical connectivity. The cortical connectivity scheme looked simple and straightforward in early years, but has turned out to be much more complex after some time and can for example be termed a distributed hierarchy (Felleman and Van Essen, 1991). Others call it a heterarchy. One can therefore assume that object recognition does not necessarily occur as strictly hierarchical but that there may be feedback interactions providing a so-called top-down influence.

Furthermore, object representations maybe encoded in a distributed manner and not by a mere single neuron. For example, Tanaka's recordings in IT showed that neighboring neurons fired for the same object, which he took as evidence that an object is distributedly represented by a local population of neurons (Tanaka, 1996) (figure 12a, 'locally distributed'). Because it seems that different visual cues are encoded in different areas, it may well be that an object is completely distributedly represented across several areas. Such a distributed representation requires a mechanism that would signal, which properties belonged together, or put differently, a mechanism that binds properties. Singer proposed that such a binding mechanism is expressed by the observed synchronization in spikes amongst far distant neurons (Singer et al., 1993) (figure 12a, 'synchronization'). This is sometimes called a *timing code*, because the specific, timed occurrences of spikes matters. Another candidate that has been proposed for binding is attention (Treisman, 1988).

Receptive Field and Firing Frequencies The receptive field (RF) of a visual neuron is roughly defined as the area of the visual field, that causes the neuron to respond with a high firing frequency (caricatured in figure 12b, 'optimal') For example, if a cortical orientation selective V1 cell is stimulated with a bar of preferred orientation, it fires at about 50 to 150Hz, depending whether the animal is anesthetized or not (e.g. (Hubel and Wiesel, 1962; Sugita, 1999)). It is now commonly accepted that such a preferred stimulus is in some sense an *optimal* stimulus - if one solely searches for a high-firing response in a neuron. Firing rates are generally lower if one adds other stimuli either inside or outside the RF. For example, if a stimulus is placed *outside* the RF, then often a suppression in the firing rate is observed (e.g. (Cavanaugh et al., 2002) for a review). Thus, one may assume that the receptive field is actually larger than when it is determined with an optimal stimulus only. The receptive field, as determined with an optimal stimulus, is therefore sometimes called the classical receptive field. Another case that can cause a decrease in firing response, is when a second stimulus is placed *inside* the RF. One possible source for such modulation in the firing rate are the horizontal connections found across hypercolumns: these connections seem to gather information from a much wider area of the visual field than the classical receptive field. Ergo, one may suspect that much more global processing takes place in area V1, than only the local analysis (Albright and Stoner, 2002; Bruce et al., 2003). The response suppression from inside the RF has been interpreted as attentional processing (Moran and Desimone, 1985).

But neuronal firing responses can be even lower, in particular when *natural* stimuli are presented (e.g. (Baddeley et al., 1997; Luck et al., 1997; Vinje and Gallant, 2000)). For example when test animals were presented a video sequence of their typical environment, V1 neurons fired at only around 14-50Hz; In humans (epileptic patients) the firing frequency of enthorhinal and temporal cortical neurons was measured around 4-10Hz for recognized images (Kreiman et al., 2000). This very low firing frequency stands in clear contrast

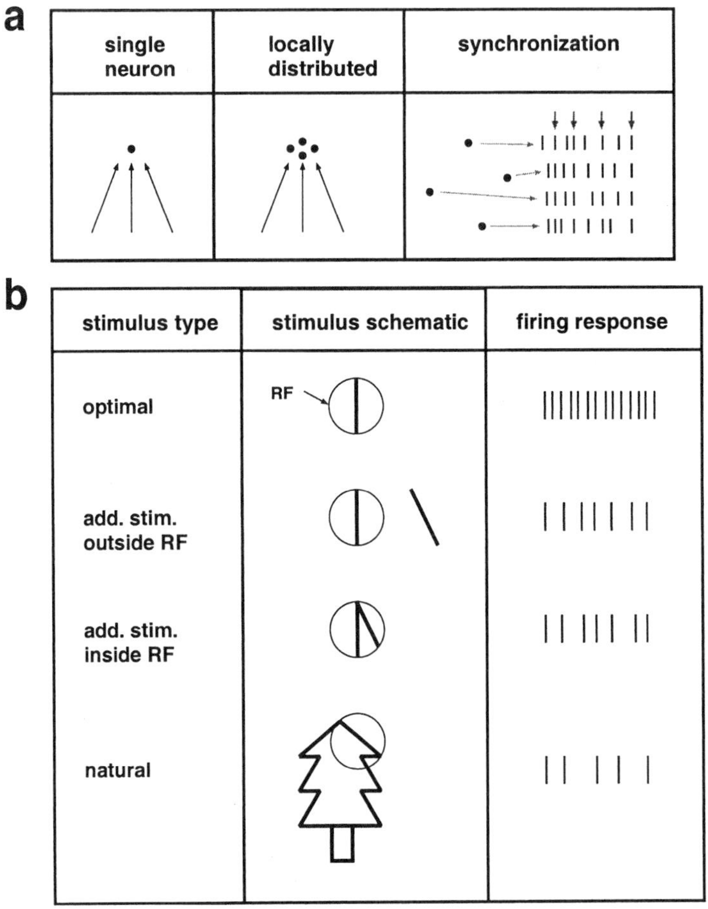

Figure 12: Coding concepts and receptive field responses. a. Some of the coding concepts existent in the computational neuroscience discipline: Single neuron, locally distributed, synchronization (or timing code). b. Response of a visual neuron to different features around its receptive field (RF): optimal stimulus, additional stimulus outside and inside the RF, natural stimulus.

with the observed high-frequency measurements of simple stimuli and one may therefore raise suspicion about the rate code and receptive field concept.

Filter Theory Some vision researchers interpret the response of visual neurons differently: Instead of interpreting these cells as detectors of contour features, they propose that these cells may filter the *spatial frequency* of the visual image and thereby perform some sort of Fourier-like analysis. This view was propelled after it was discovered that the response characteristics of V1 cells are more subtle, when they are stimulated with sinusoidal gratings - instead of a single line or bar only (De Valois and De Valois, 1988). For example, the orientation tuning curve looked as shown in figure 10d. The horizontal dashed line crossing the orientation tuning curve represents the 'spontaneous' firing rate, that is the mean firing rate when a neuron is not stimulated. This spontaneous firing rate can be in the range of less than one spike per second (0.5 Hertz) to up to several spikes a second. For the preferred grating orientation the cell fires with a high frequency - as recorded in experiments using single line orientations. But for grating angles that deviate a few degrees off the preferred orientation, the response of the neuron is actually below the spontaneous firing rate. A second 'refined' finding was, that the receptive field profile looked more like a Gabor[1] function than a Gaussian (not shown). A third intricate discovery was that different cells showed a preference for gratings of specific frequency. Taken together, it looks like there are cells that code for different spatial (or resolutional) scales and for different orientations, which suggests that some sort of wavelet encoding may occur in the primary visual cortex. If one translated that into the picture of the orientation columns, then one would add the dimension of spatial scale to each orientation (figure 10e). Many vision models perform image-filtering operations inspired by the above framework. They use filters whose exact form is motivated by those recordings. For example, an elegant model by Perona and Malik performs texture segregation and produces a contour image as output (Malik and Perona, 1990). A number of neurophysiology studies have tried to find filter detectors in higher areas (V2 and upwards in the cortical hierarchy). They used polar and hyperbolic patterns that could possibly serve as complex filter detectors (e.g. (Hegde and Essen, 2000; Gallant et al., 1996)). Such patterns are shown in the bottom row of figure 11.

3.3 Speed

Latency Code Many of the neurophysiological experiments are carried out using (visual) presentation times of several hundreds of milliseconds, which is a long time span compared to the fast-paced dynamics of visual selection: saccades are launched every 200-300ms, attentional shifts are carried out several times between saccades (Parasuraman, 1998). One may therefore question the interpretation of experiments with long stimulation durations. Indeed, it has long been neglected that visual processing occurs blazingly fast. Although Potter had already shown this with behavioral experiments (see section

---

[1] A Gaussian multiplied by a sinusoid (Gabor, 1946)

1.2), it was Thorpe who sharply pointed out this characteristic using event-related potential (ERP) studies (Thorpe et al., 1996) (see also (Schendan et al., 1998)). In his experiments, subjects had to decide whether a picture contained an animal or not. The picture was presented for only 20ms. The analysis of the ERP patterns showed, that after 150ms only, a subject has already made a reliable decision and that this decision was made in the frontal cortex, indicating that visual information has made the loop from V1 over IT to frontal cortex somehow. It should be mentioned however, that Thorpe's and Potter's experiments work with expectation: the subject knows in about what to look for - in other terms certain frames may have been activated already and this preactivation could possibly reduce the reaction time by some amount. Still, 150ms is blazingly fast and without expectation it would probably take only a few tens of milliseconds more.

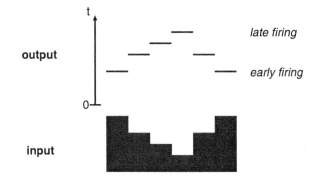

Figure 13: A latency code. The amount of input determines the onset of firing: a large input triggers early onset, a small input triggers late onset (0: onset of stimulus, t: time). This type of timing code would swiftly convert a complex pattern into a subtly timed spike pattern. Proposed in variants by Thorpe and Hopfield. Adapted from Thorpe, 1990.

Because visual processing happens so rapidly - and possibly any neural processing -, one may doubt whether there is a frequency code at work, because such a code may be simply too slow. Thorpe himself therefore proposed a timing code in which computation is performed by precisely timed spikes (Thorpe, 1990). Figure 13 expresses the idea. The amount of input determines when the cell fires its first spike after stimulus onset: a large input triggers early firing, a small input triggers late firing. VanRullen et al. developed a network that is made of a large stack of neuronal layers using this type of encoding. The network is designed to mimic Thorpe's visual experiment (Rullen et al., 1998). The idea of such a latency code has also been proposed by Hopfield (Hopfield, 1995).

Global-to-local Some psychologists were aware of the the blazing processing speed long before the above debates and were wondering how the visual system may analyze visual information so rapidly. They argue, that because we can recognize the gist of a scene so rapidly, that the visual system processes a scene by decomposing it in a global-to-local manner, basically the reverse to the neuroscientific paradigm. Neisser has formulated this idea in the 60's already (Neisser, 1967), Navon was the first to seriously investigate this concept (Navon, 1977). Navon described this view concisely with *'forest before trees'*: In experiments with large letters made of small letters (figure 14), Navon tried to prove that first a global perception of the large letter takes place, followed by a local perception of the small letters. If such a 'global-first'

```
S           S
S           S
S           S
S           S
S S S S     S
S           S
S           S
S           S
S           S
```

Figure 14: Navon's task to test for a global-to-local recognition evolvement: Is the global 'H' or the local 'S' perceived first?

evolvement would take place in the visual system, one may wonder how. Because it had to happen fast, global processing may have to take place in low cortical areas already. This stands in apparent contrast to the long-assumed picture that only a spatially local analysis occurs in V1. But this picture has already been criticized from two sides. One side are the neurophysiological recordings on the characteristics of the receptive field (previous section). Another side are the psychophysical measurements on contour integration, which evidence that perceptual grouping operations may take place in V1 (Kovacs, 1996; Hess and Field, 1999). It may therefore be worth considering whether a possible global-to-local analysis starts in V1 already. On a complementary note, there is an attempt to model a global-to-local analysis (see (GERRISSEN, 1984; GERRISSEN, 1982)).

3.4 Alternative 'Codes'

The dominant view of how neurons communicate with each other is guided by regarding it as some sort of Morse code or also called a 'neural code' (Koch, 1999). The sequence of spikes, that a neuron generates, is interpreted as being part of a rate or timing code. Both, rate and timing codes, come in many different variants (deCharms and Zador, 2000). We have mentioned a few examples previously. An alternative to this Morse code thinking is to regard spikes merely as connectors between far distant sites: the sequence of spikes would have no meaning, but each single spike may be part of a computation connecting distal sites.

Cortical Potential Distributions   One such alternative is Tuckwell's theory, in which the critical measure is a potential distribution across cortex (Tuckwell, 2000). One component of the cortical potential distribution can be the *field potential*, which is a voltage that reflects primarily post-synaptic potentials and only to a minor extent the spikes themselves. Another component can be the *magnetic fields*. Both potentials can be regarded as a general electromagnetic field description for cortical states. According to Tuckwell, such global cortical potential distributions could define cognitive states. A change between these states could occur very rapidly simply by the simultaneous change of the entire distribution at each locus.

Waves   Another alternative would be to regard an observed (or measured) spike as being part of a traveling wave. Traveling waves exist in the nervous system of many animals (e.g. (Hughes, 1995; Prechtl et al., 1997; Wilson et al., 2001; Shevelev and Tsicalov, 1997). Generally, they are considered as non-functional, accidentally emerging from networks of neurons. Recently, there has been some effort to attribute computational functions to traveling waves. Jacobs and Werblin measured traveling waves in the salamander retina in response to visual stimulation, and speculated that it may have edge enhancing effects (Jacobs and Werblin, 1998). They envision that such waves are possibly involved in different neural computations. Barch and Glaser use excitable membranes to detect motion: different motion signals leave traveling waves of characteristic shape on an excitable membrane (Barch and Glaser, 2002).

Yet, long before the discovery and reinterpretation of traveling waves, there existed some computational reflections on coding and visual representations, which have never really been pursued in neuroscience. One is the broadcast receiver principle and the other is the idea of self-interacting shape. Both use the idea of wave propagation, which can be readily interpreted as traveling waves.

Broadcast Receiver Principle   This idea has been expressed by several people (e.g. (Blum, 1967; Deutsch, 1962), see (Blum, 1967) for references). Blum has given the most illustrative example, see figure 15a. The schematic depicts an excitable 3D volume that propagates waves. He imagined that a feature extraction process places filtered properties (of a visual field for example) onto this *propagation medium.*

Wherever these triggered waves would meet simultaneously, the neuron at that locus would signal the presence of those properties. Blum
called this the simultaneity observer, which in modern terminology
would be called coincidence detector. In some sense it is an integration by propagation, as opposed to for example the integration by
convergence or synchrony (figure 12a). The advantage of integration
by propagation is that it is not bound to a specific wiring pattern.
Blum suggested this process in order to address the problem of 'fast
access'. He considered this, as Thorpe does recently, as the foremost
issue that needs to be addressed in visual recognition.

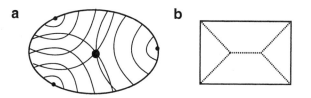

Figure 15: Coding by waves. a. Broadcast receiver principle: 3 features have been extracted and their signal is placed on an excitable 3D
medium that propagates waves. At the location where all three waves
coincide, integration would take place. b. Self-interacting shape.
The symmetric-axis transform: the rectangle has been transformed
into the dotted symmetric-axes by contour propagation (more on this
transform in chapter 7). Adapted from Blum 1967, 1973.

3.5 Alternative Shape Recognition

Visual shape is generally described by its contours, like in the computer vision approaches mentioned in chapter 2 or like in the neuroscientific feature integration concept. An alternative is to encode the
space that a shape engulfs. One method to achieve that would be by
extracting the spatial frequencies like in the channel theory. Another
method would be to encode the region directly: early Gestaltists proposed that a shape interacts with itself, (e.g. (Koffka, 1935)). During
such a self-interaction, the enclosed space is automatically encoded.
Based on these ideas there have been several models (see (Blum, 1967;
Psotka, 1978) for references). We discuss two of them.

   The most specific and influential one is probably Blum's symmetric-
axis transform (SAT) (Blum, 1973) (see figure 15b). He specifically designed the transformation to perform as a biologically plausible process. The idea is to dip a shape into an excitable medium that would
trigger a 'grassfire' process as Blum called it. A more modern term for
grassfire would be *contour propagation* or *wave propagation.* Wherever these propagating contours collide, they cancel each other out,

evolving a geometrical pattern in time called the symmetric axis.

There is some psychological and neuroscientific evidence, that the SAT is carried out in the primate visual system. For example, Psotka asked human subjects to place a single dot into a shape. It turned out, that the superposition of all the dots (from many different subjects) is a pattern that resembles the symmetric axis for that shape (Psotka, 1978). Kovacs and Julesz found that human sensitivity to orientation stimuli is higher in the center of a region, which they interpreted as evidence for SAT-like processes (Kovacs and Julesz, 1994). Kovacs in particular summarizes a number of psychological studies that support that regions are transformed into 'skeletonal' structures in V1 already (Kovacs, 1996) (see also 'Global-to-local' in previous section). A neuroscientific study on primates indicates, that the SAT or similar region-based processes, may take place in the primary visual cortex (Lee et al., 1998).

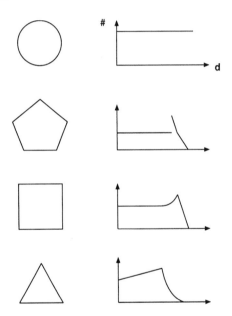

Figure 16: Shape description by a distance measure. Deutsch suggested a mechanism that measures the distance between pairs of points along the contour. Four examples are shown. The histogram plots the number of 'distance pairs' (#) for a given distance (d). Redrawn from Deutsch 1962.

Another type of shape interaction was suggested by Deutsch (1962). Based on the histology of the bee's optical lobe, he contrived a system that would 'measure the distance between all pairs of points on the boundary of the shape and so produce a distribution of how many pairs of points there are at each distance from another', see figure 16 for examples. He posited that a shape recognition system should possess qualia that allow the system to recognize the shape independent of its position, size, orientation and mirror image. His shape encoding mechanism could account for all of these aspects except of size invariance. We discuss some of these aspects later again in chapter 9.

3.6   Insight from Cases of Visual Agnosia

The study of humans with visual deficits may give us also some insight on the functioning of the vision process (Humphreys and Riddoch, 1987a; Humphreys and Riddoch, 1987b; Farah, 1990). It is a study on the behavioral level again and the following observations may also have fitted into chapter 1, but we felt that they are better placed in the neuroscience chapter. Drawing conclusions from agnosia cases is especially intriguing, because many of these cases can appear very similar at a first glance, but are actually very individual in their deficits. Humphrey and Riddoch (Humphreys and Riddoch, 1987a; Humphreys and Riddoch, 1987b) concentrated on a few subjects and analyzed them thoroughly with a large variety of tasks. For example, subjects were tested with one of the tasks that Navon used to determine whether perception may occur in a global-to-local manner (see section 3.3). Some subjects struggled with some of these letter tasks, for example some subjects were able to comprehend either the global or local letter only. One subject was able to recognize the outlines (silhouettes) of line-drawing objects better than if they were presented the same object with all its major contours. From such observations, Humphrey and Riddoch concluded that two separate structure-processing streams exist, one for local feature processing and the other for global shape processing (figure 17). Both of these streams converge to viewpoint-dependent object descriptions and to abstract episodic object descriptions, which are collections of viewpoint-dependent descriptions. Both object descriptions point towards object form knowledge, which in turn points toward semantic knowledge.

But recognition evolvement maybe even more individual than just a separation into a local and a global structure-processing stream. To point this out, we mention a few cases which Farah has described (Farah, 1990) and we thereby follow her classification, because it is the most recent attempt to classify those multi-faceted cases:

1) One subject who suffers from 'apperceptive agnosia' had greater difficulty with curved lines than with straight lines, which evidences that there may be distinct mechanism to evolve curves and straight lines.

2) Some of the *associative* subjects - classified as 'narrow sense'

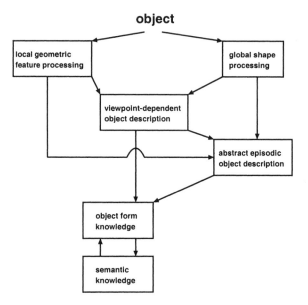

Figure 17: Humphrey and Riddoch's scheme of object recognition as derived from studies of visual agnosia cases. Redrawn from Humphrey and Riddoch, 1987a.

by Farah - were able to 'see'[2] line-drawings and to slowly copy them (figure 18). For example a structure like in the top left (under 'shown'), was copied in a part-wise manner, meaning it has been perceived as two separate regions, see middle, schematic graph under 'patient's copy'. An alternative may have been that the structure was copied as shown under 'alternate possibility'. One may conclude from this example, that because the patient copied the drawing region-wise, that perception is based on parts or regions. The second example shows that the patient copied a wheel with spokes as a circle with lines going through it (bottom row in figure 18). An alternative could have been that the subject copies the structure slice-wise, which would be a region-wise manner. This example is in contrast to the first one. It does not point toward a perception based on regions but to a perception of spatially distinct features.

3) Quite a number of visual agnosia cases have troubles recognizing objects of certain categories, in particular subjects that have difficulties with face recognition. Those subjects could not recognize for example fruits and vegetables, or objects found indoors. This may evidence a failure of making an association to 'action representations'. Still, one may also take this as evidence that object representations

---

[2]but not recognize

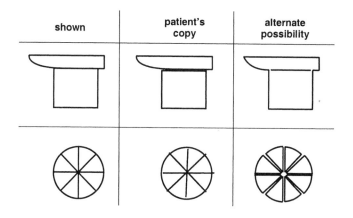

Figure 18: Some of the copying results (schematic) of an associative case (narrow sense). Left: what the subject was shown. Middle: the way the subject copied it. Right: theoretically, alternate possibility.

are spatially very distinct.

In summary, we regard these findings as an indication that perception - or recognition evolvement and representation - may be more individual to an object and does not follow a standard evolvement scheme.

3.7   Neuronal Level

Until now we have only addressed the system level. But what may we learn from the neuronal level? We start with the electrical properties of the nerve cell membrane.

Subthreshold dynamics   Below the spiking threshold, the membrane behaves roughly like a *resistor-capacitor (RC) circuit*, in which the capacitance, C_m, represents the membrane capacitance; the resistor, Rleak, expresses the 'leakiness' of the membrane; and the voltage node, V_m, represents the membrane voltage (figure 19a). If one stimulates the nerve cell membrane with a step current of low amplitude (see two 'I' traces), its membrane voltage will gradually increase and then saturate. After offset of the stimulating current, the membrane voltage will bounce back (see corresponding two 'Vm' traces). How fast this increase and decay occurs depends on the parameter values for the resistance and membrane capacitance. A neuron's membrane capacitance can be reliably estimated, but its membrane resistance is not alway easy to determine, especially in case of the small neocortical neurons. The decay times range from a few milliseconds to a couple of tens of milliseconds (Johnston and Wu, 1995). If the decay time is rather long, a neuron may function simply as an *integrator*, because it would continuously add up synaptic input. In contrast, if the decay time is short, then the neuron may function as a coincidence detector: synaptic input is only added up, when it occurs almost simultaneously, otherwise synaptic input leaks away before other synaptic input adds up. In which one of these two modi a cortical neuron operates, is still an on-going debate (e.g.(Koenig et al., 1996; Koch, 1999)).

Synaptics The synaptic responses can be very diverse. There exist the simple excitatory and inhibitory synapses, that cause a transient increase or decrease in the membrane potential respectively. A schematic *excitatory postsynaptic potential* (EPSP) is shown in figure 19b on the left side. Its duration lasts several milliseconds. Two excitatory synapses sum up linearly as shown in figure 19b (upper right). There exist also voltage-dependent synapses that produce non-linear summation: they generate a larger synaptic response, when the membrane potential has already been elevated by other stimulation (figure 19b, lower right). This type of response can be interpreted as multiplicative (gray trace indicates mere summation). There are a number of other synaptic interactions, as for example divisive inhibition. This variety of synaptic responses could therefore perform quite a spectrum of arithmetic operations (Koch, 1999).

Dendritics The tree-like dendrite of a neuron collects synaptic signals in some way. In a first instance, the dendrite may be regarded as a *passive cable* conveying the synaptic signals that were placed on it. The propagation properties of such a dendritic cable have been characterized to a great extent by Rall (Rall, 1964). For practical simulation purposes, a dendritic tree is discretized into so-called compartments, each of which simulates a patch of membrane by a RC circuit. The compartments are connected by a horizontal (or axial) resistance, Raxial, which simulates the internal, cellular resistance. Figure 19c (left) shows two compartments connected by a horizontal resistance. If synaptic input is placed for example in the left compartment (voltage node Vm1), then it will spread towards its neighboring compartments and decay in amplitude due to the horizontal resistance and the membrane resistance of each RC circuit. The schematic in figure 19c (right side) shows this propagation decay for a synaptic input placed on Vm1. Given these decay properties, a dendrite may be regarded as a low-pass filter. But there maybe more complex processing going on in a dendrite than just mere synaptic response integration along a passive cable. Because many dendrites contain also voltage-dependent processes that could cause quite a number of different operations. For instance, a dendritic tree sprinkled with the voltage-dependent 'multiplicative' synapse just mentioned previously, could carry out sophisticated computations (Mel, 1993). Or a den-

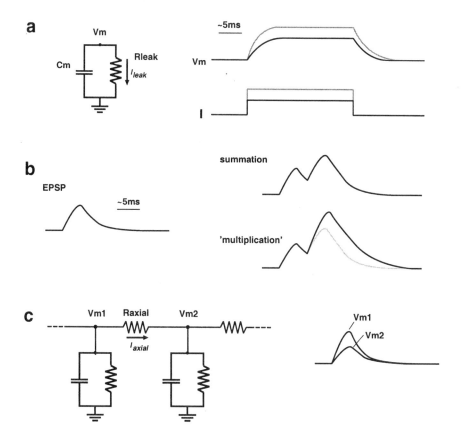

Figure 19: Schematic neuronal dynamics. a. Resistor-capacitor dynamics represent the nerve cell membrane dynamics (below spiking threshold). Left: RC circuit: Cm: membrane capacitance; Vm membrane potential: Rleak: membrane resistance. Right: The response of Vm to two (superimposed) step currents. b. Synaptic response and integration. Left: typical excitatory postsynaptic potential. Right: linear summation (above), multiplicative summation (below). c. Compartmental modeling of a piece of dendritic cable. Left: A cable is well simulated by discretizing it into separate RC circuits connected by axial ('horizontal') resistors, Raxial; Right: Synaptic response decay along compartments.

dritic tree containing voltage-dependent channels like the ones in the soma (see next paragraph) could actively propagate a synaptic input toward the soma. Furthermore, spike waves have been discovered that run up and down the dendrite (e.g (Svoboda et al., 1997). Thus, there could be a substantial amount of *active* processing taking place in a dendrite (Johnston et al., 1996; Koch, 1999).

Somatics (above threshold dynamics) In the soma, spikes are generated. Once the membrane voltage crosses a certain threshold, an action potential is generated that is the result of a marvelous interaction of subtly timed ionic currents (Hodgkin and Huxley, 1952). Such an impulse is generated within 1-2 milliseconds. If the neuron is stimulated continuously, for example with a step current of sufficient magnitude, then the neuron spikes repetitively. There is a large number of models that mimic this spiking behavior ranging from detailed models, that simulate the exact dynamics of the ionic current, over to oscillatory models and to simplified spiking models. From this wide range of suggested neuronal functions and models we here mention only one type, the integrate-and-fire (I&F) model (Koch, 1999). It is the simplest type of spiking neuron that incorporates explicit subthreshold and spiking dynamics.

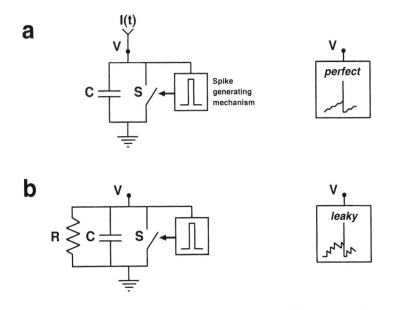

Figure 20: Integrate-and-fire (I&F) variants. a. Perfect I&F neuron: C represents the membrane capacitance, V the membrane potential, the box with the pulse icon represents the spiking unit which activates a switch (S) resetting the membrane potential. b. Leaky I&F neuron: it contains a resistance (R) in addition. Modified from Koch, 1999.

The most elementary I&F variant is the *perfect* I&F neuron (figure 20a). It merely consists of a capacitance, representing the membrane capacitance, and a spike-generating unit possessing a threshold. When synaptic input reaches the (spiking) threshold, the spike-generating unit will trigger a spike and reset the activity level back to 0 (ground). If desired, a refractory period can be simulated by clamping the activity level to 0 for a short while. This model just suffices to simulate simple synaptic integration. A more elaborate variant is the *leaky* I&F neuron (figure 20b), which consists of the perfect I&F circuit plus a membrane resistor. Or put with above terminology, it is the RC circuit plus a spiking unit. The membrane resistance has the effect that any input is only transient, which renders this model much more realistic than the perfect I&F model. This model is for example suitable for detecting coincidences amongst synaptic input.

3.8 Recapitulation and Conclusion

The experimental results in visual systems neuroscience are fascinating but remain puzzling as a whole. The innumerous recordings and measurements have been interpreted in many different ways and have not led to a unifying picture of the recognition process but rather to a continuing diversification in theories and approaches. The prevailing idea of feature integrating does not directly help us out in specifying the loose category representations we look for and that is primarily for two reasons:

1) If the features found in the visual field are integrated, then a single category instance has been described, an object of the identity level in some sense (figure 3). What would have to follow that description, is a continued evolvement leading to the abstract category representation, which is able to deal with the structural variability.

2) Most of the feature or object detectors found in higher areas do not exactly represent structure seen in the real world. The exception are a few neurons firing for real-world objects, yet they hardly represent abstract representations but rather objects of the identity level (figure 3). The feature integration concept would be much more compelling if there were neurons that fired for different instances of the same category, for example a cell responding to structurally different trees, fruits or flowers.

The neural networks performing this feature integration do not make an explicit distinction between representation and evolvement - as opposed to the computer vision systems reviewed in the previous chapter. Evolvement is representation and vice versa. The strength of many of these neural network architectures is that they learn automatically after some initial help. Their downside is that their neuronal units or connections share too many object representations and this shared substrate leads to a considerable overlap. Thus, they are structurally not specific enough to represent different categories. In order to differentiate between structurally more diverse objects, a network has to evolve structurally more distinct features and contain distinct representations. The cases of visual agnosia suggest that the visual system may indeed have mechanisms that perceive specific structures and that object representations maybe spatially very distinct.

The low spiking frequency found with complex stimuli presentations stands in apparent contrast with the lightning processing speed and casts serious doubts on the prevailing rate-code concept. Several researchers have therefore looked for alternative codes, ranging from wave propagation, over to timing codes, to cortical potential distributions. As we have already pointed out in chapter 1, we also consider speed a pivotal characteristic that needs to be explained when one attempts to unravel the evolvement and representation of the visual recognition process - and likely any computation in the nervous system.

If one adopts the idea that waves perform computation - as we will do later - then one may regard a spike merely as part of a wave running across the neuron. In order to account for the processing speed, such a wave would need to be fast. We argue in our later chapters - in particular chapter 7 - that fast waves may exist (section 7.5). Taking the 'wave view', the occasionally observed high firing frequencies (for simple stimuli) may then be interpreted as the continuous accumulation (or collision) of waves, because the visual system is idling due to the lack of meaningful input.

The detailed function of the neuron remains also an enigma, despite quite a number of experiments and model simulations: Is it a simple integrator? A sophisticated calculator? A coincidence detector? Is its dendrite a fine-tuned 'spatio-temporal' filter? In the remaining chapters we will only make use of the 'crude' I&F neuron type, not because we think that is what the neuron can be reduced to, but because it suffices for the wave-propagation processes we will describe in later chapters.

4 Neuromorphic Tools

The operation of the nervous system requires amazingly little energy. The average nerve cells dissipates power in the 10^{-12}-watt range; the average logic gate in a computer dissipates 10 million times as much (Mead, 1989). Can one not create electronic circuits that operate as *swift and economic* as the nervous system's circuits do? Neuromorphic engineers do so using *analog circuits*, whose graded, analog voltages and currents are the computing signals, and whose emulation runs in real-time and with little energy consumption (Douglas et al., 1995). The idea was started in the 60's using bulky electrical devices very much like the ones in (old) radios and TVs, but it was soon flattened by the uprise of digital integrated circuits. The idea was effectively revived in the late 80's by Mead, who built analog circuits using modern chip technology, so-called very-large-scale-integrated (VLSI) circuits (Mead, 1989). The most famous example of such an analog VLSI (aVLSI) circuit is the 'silicon' retina, which simulates some of the processing taking place in the retina (Mahowald and Mead, 1991). In this chapter we present a few simple circuits that simulate those neuronal dynamics we have just sketched in the last section (3.7) of the previous chapter.

4.1 The Transistor

To grasp the gist of the neuromorphic approach, it is convenient to introduce the digital operation of the transistor first. A transistor has a gate, drain, source and a channel connecting source and drain (figure 21). The voltage at the gate determines the amount of current flowing through the channel. In a digital circuit, the gate voltage is either 0 or 5 Volts, corresponding to the binary values 0 and 1. The corresponding current flowing through the channel is therefore either

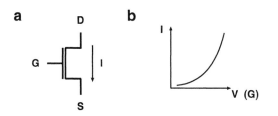

Figure 21: Transistor schematic and analog electrical properties. a. Icon of a transistor. D: drain, G: gate, S: source, I: current. The amount of current flowing through the channel is determined by the gate voltage. b. Current-voltage relationship in the analog domain: it is exponential in the subthreshold domain.

fully on or totally off. In an analog circuit in contrast, the gate voltage is typically in a low voltage range from ca. 0.1 to 0.8 Volts generating small currents in the range of pico and nano ampere. In this so-called *subthreshold range* the amount of current flowing through the channel is an exponential function of the applied gate voltage. Using this current-voltage relationship, one can construct other useful functions, like a sigmoid, which would be generated by a circuit called the transconductance amplifier. For the present essay, it suffices to understand that a small gate voltage will generate only small currents.

4.2 A Synaptic Circuit

A simple postsynaptic response, as schematically shown in figure 19b, can be generated with a circuit consisting of only *four transistors* (figure 22a). Transistor T1 receives presynaptic (binary) voltage spikes and therefore operates pretty much like a transistor in digital circuits: on or off. Transistor T2 emulates the weight of the synaptic circuit: at its gate, a small voltage is applied, which reduces the large 'digital' current flowing through transistor T1, to a smaller current IWGT. This 'weighted' synaptic current then makes a quasi-U-turn through transistors T3 and T4, in which the sign of the current - pre-

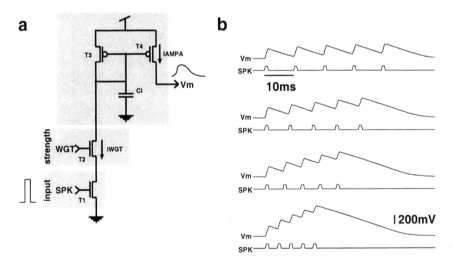

Figure 22: Synaptic circuit and response. a. Circuit made of four transistors (T1-T4) and a capacitance (C1). Input: (presynaptic) voltage pulse, SPK, at T1. Output: postsynaptic-like response, Vm. b. Response of a silicon synapse (Vm) to presynaptic stimulation (SPK). Four different spike frequency stimulations are shown (the presynaptic spikes are scaled down). From Rasche and Douglas, 1999.

viously negative - is inverted and then dumped as a positive current, IAMPA, onto a capacitor emulating the neuron's membrane potential (capacitor not shown in figure). The capacitance C1 delays the synaptic current to make it resemble a real postsynaptic current: without that capacitance, the synaptic current would be pulse shaped like the pulse-shaped presynaptic spike entering the circuit at the gate of transistor T1. Figure 22b shows the synaptic response for a presynaptic stimulation of five spikes for four different stimulation frequencies.

This silicon synapse only outlines a raw circuit. Many little amendments can be made to convert it for example into a learning synapse (e.g. (Häfliger and Rasche, 1999)), an inhibitory synapse (e.g. (Rasche, 1999), a voltage-dependent synapse (Rasche and Douglas, 1999) and a depressing synapse (Rasche and Hahnloser, 2001).

4.3 Dendritic Compartments

Cable-like signal propagation can also be emulated in aVLSI. One meets a fundamental problem though, the implementation of a useful resistor in analog circuits. There exists no actual resistor device, but a lot of work-arounds. Depending on the task, elegant solutions can be found (Mead, 1989), but for the emulation of dendritic signal spread the lack of useful resistors is even more blatant. One method is to use so-called switched capacitors, which consist of two transistors aligned sequentially and a small capacitance between them (figure 23a). The alternate digital activation of the two transistors by a clock, piecewise moves the charge from one side of the resistance to the other.

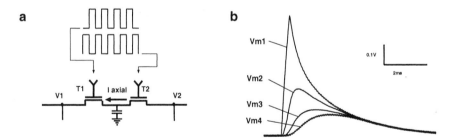

Figure 23: Dendritic propagation in aVLSI. a. Emulating a resistor using switched capacitors: two transistors (T1, T2) in sequence with a tiny capacitance between the transistors. The transistors are driven alternatively by a clock. b. Propagation of a postsynaptic signal along three compartments (postsynaptic signal placed in compartment 1 [Vm1]). From Rasche and Douglas, 2001.

The size of the resistance would depend on the switching frequency - the speed of the clock - and the capacitance between them. The drawback of this method is the use of clocks, which renders the approach rather a mixed analog-digital implementation. Elias has modeled dendritic cable properties using this type of resistance for both resistances - horizontal and leakage - of the compartmental modeling approach (Elias, 1993). We also have played around with this method, but preferred to simulate the leakage resistance with either an amplifier or a simple transistor that drains activity continuously. Figure 23b shows the propagation decay for a synaptic signal running down three compartments (Rasche and Douglas, 2001).

4.4   An Integrate-and-Fire Neuron

The I&F neuron model can be emulated in analog VLSI with a mixture of analog operating transistors and some digital circuitry for generating the neuronal spike (figure 24). A schematic version of the elementary spike generating circuit is shown in figure 24b: it consists of two capacitances, C1 and C2, and an amplifier A. If the membrane node voltage, Vm, crosses a certain threshold, for instance by current injection (Iinj), the amplifier will generate a sudden strong current and cause Vm as well as node Vspk to rise abruptly. Thus, there are two separate dynamic variables, Vm, the actual membrane potential as well as Vspk, a digital output of the circuit. Figure 24c shows the dynamics of these two nodes. The spiking threshold is denoted as Vt in the upper graph. Once Vspk is high, it will turn on M1 (lower left transistor in figure 24a) and draw a current, IReset, from the membrane node via the transistor with gate voltage RESET. These two transistors act like the bottom two transistors (T1 and T2) of the synaptic circuit (in figure 22). It is therefore the gate voltage RESET that determines how fast the spike resets, which is expressed by the time period *th* in the lower plot of figure 24c.

The amplifier consists of four transistors, which are basically two 'NOT' gates (digital circuitry) aligned in sequence. The gate voltage PU of the upper left transistor determines the spiking threshold.

This circuit represents a perfect I&F integrator and can be modified to emulate the various I&F types discussed in the previous chapter. For example, to make it a leaky neuron, one would simply add another transistor that continuously drains activity away from the membrane node.

4.5   A Silicon Cortex

The amount of circuitry that can be stuffed on a single chip is limited. To put it more direct: only a limited number of silicon neurons can fit on a single chip. It is therefore necessary to distribute the circuitry over several chips, for example having a retina circuit on one chip, a part of the cortical circuit on another chip and so on. This naturally

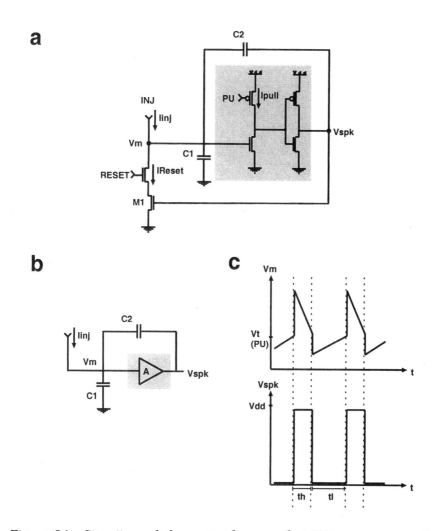

Figure 24:   Circuits and dynamics for a perfect I&F neuron.   a.   6 transistors and 2 capacitances.   b.   Schematic of the spiking unit (2 capacitances and amplifier).   c.   Dynamics of the Vm and Vspk node. Adapted from Mead, 1989.

leads to an architecture that resembles the real visual system in some sense. The neuromorphic term for such a system is *multi-chip archi-tecture*. In this neuromorphic architecture, chips communicate with each other by pulses, very much like the computational maps in the biological visual system do, e.g. the retina sends spikes to the orien-tation columns. The challenge is to get the chips to fluently communi-cate with each other. This problem does not appear for the operation of digital circuits, because they are synchronized by a common clock determining their pace. In contrast, the operation of the analog multi-chip system needs an asynchronous communication scheme allowing for an exchange of pulses at any time, meaning without relying on any pace maker. Such an asynchronous communication scheme has been invented by several groups (Deiss et al., 1999). I mention here only the one I grew up with, the Silicon Cortex. Originally designed by Mahowald and collaborators, it is now in the process of being refined and tested by Douglas and coworkers (Liu et al., 2001). It uses a so-called *Address-Event Representation* and basically consists of a list of the specific neuronal connections between chips, whose wiring has to be programmed.

4.6 Fabrication Vagrancies require Simplest Models

There is a little catch though with the emulation of analog circuits. Because the fabrication process of chips does not generate each transistor exactly equally - irrespective whether it contains digital or analog circuits -, there exist slight variations amongst transistors: even if one applies the exact same gate voltage to different transistors on the same chip, the resulting currents flowing through the channels are slightly different which can cause inequalities in summation. These variations do not pose a problem for digital circuits, because they operate in those two extreme states only. Due to this *fabrication noise*, models that are intended for aVLSI construction, need to be simple and robust. There are a number of simple tricks to counteract to these fabrication variations, for example to increase the transistor size for crucial transistors, like the weight transistor in a synaptic circuit, or to incorporate adaptive circuits into the design like in the silicon retina. Still, simplicity and robustness must accompany any approach to neural modeling in analog circuits.

4.7 Recapitulation

Almost anything goes in analog VLSI circuits - as long as it is simple and robust. One can emulate a palette of synaptic responses, the propagation qualities of dendritic cables and a variety of somatic spiking patterns. A multi-chip architecture enables the communication between several chips and so allows for the distributed emulation of large networks. The really exciting part of this entire approach is, that it can run at reasonable low power, compared to a pure digital approach. Moreover, it is blazingly fast: it runs in real-time like the real visual system does.

Neuromorphic engineering goes of course way beyond of what we have touched in the few paragraphs: there exist silicon cochleas, olfaction systems, visual attention systems and much more (e.g. (Mead,

1989; Indiveri, 2001)): The few above mentioned circuits suffice to envision an implementation of the networks we will be talking about in some of the remaining chapters (6-10).

5 Insight From Line Drawings Studies

We now take a first step toward exploring the possible nature of category representations. We do this using line drawing objects and computer vision methods. The line drawings represent only the rough structure of objects (see figure 25). They are certainly a simplification of the existent variability in real-world objects, but we will show in chapter 7 how we can apply the experiences and discoveries made in the present chapter to real-world objects depicted in gray-scale images.

The studied categories are chair, desk, bed, table, drawers and closet. The detailed voluminous structure of each single part, like the voluminous shape of a chair leg, is omitted, but can be imagined to be there. We chose the following two main restrictions on part-shape variability: 1) Only straight lines are used. 2) Only rectangular surfaces are used. The drawn objects show part-alignment variability and sometimes part redundancy. The objects are shown primarily in canonical views, yet not from fixed viewpoints. The *goal* is to find a set of 2D descriptors which is easily extractable with a few rules. The type of recognition system we pursue is therefore a 2D-2D recognition system (figure 8 in chapter 2). Because we are looking for loose representations, which we have not defined exactly yet, we approach this task as follows: we attempt to get the *categories distinguished* with as few features as possible.

Initial Thoughts Let us first think through a hypothetical description scenario. Imagine one had to specify these objects using only line pieces as elementary features. A chair would have to be described as three to four lines with an approximate vertical orientation and some spacing in between them - representing the legs -, another few lines with approximate horizontal orientation connecting to a surface - representing the seating area -, and so on. Such a description would result in an overwhelmingly large amount of structural relations between lines. It therefore makes sense to extract higher features first, like vertex features (intersections of two or more lines) or surfaces, with which one could form representations having less structural relations. That is what we do in the first subsection.

5.1 A Representation with Polygons

In a first step, L features are extracted (2 intersecting lines). In a second step, these L features are used to form two basic features, the surface rectangle and the '3-line-polygon' (figure 26a, 1 and 2). The surface rectangle represents surfaces that are tilted and slanted in space like rectangles, trapezoids and parallelograms. The 3-line polygon consists of three sequentially connected lines and can outline any

Figure 25: Categories used in the line-drawing studies. Left side: chair (top), desktop (middle), drawers (bottom). Right side: bed (top), table (bottom).

structural characteristic of a category. An alternative to these surface and 3-line polygon features would be vertex features made of more than 3 intersecting lines, but such features are rare in objects of our environment and are difficult to extract. Fu has already pointed that out: this is the reason why he chose to describe his objects by surfaces (Lee and Fu, 1983). The surface rectangle is then used to form features like nested surfaces, folded surfaces and parallel surfaces (figure 26a, 3 to 5). None of the structural relations amongst these surfaces has been specified exactly in order to be able to deal with the part-alignment variability.

With this set of extracted features, category representations are formulated. For example a chair is represented by two different 3-line polygons and a surface (figure 26b, chair, 'entire rep'). One 3-line polygon is formed by the frontal legs of the chair and the connecting seating contour: it is a U-like feature whose outer lines point downwards, which we call a 'bridge'. As the chairs legs are sometimes askew, we define the direction of the bridge's legs loosely with a tolerant angle (see α in figure 26b, chair, 'bridge'). Because the bridge feature per se is not specific enough for chairs, it is also found in other categories as well. We therefore combine two bridges to a 'double bridge', sharing a close or even common leg. The middle angle of both bridges should differ by a minimum angle (see α in figure 26b, chair, 'double-bridge'). Generally, such double bridges are characteristic to objects having legs as part of their structure and are thus also found in tables and beds. The second characteristic 3-line polygon is the 'seat': A chair's leg the seating contour and the back-rest contour form a Z-like feature (figure 26b, 'seat'). The surface is taken to be the back-rest. Because the back-rest can be askew as well, the surface rectangle geometry was not defined exactly. To form the category representation, the three features - double bridge, seat and surface rectangle - were set only into loose relation.

A desktop is generally made of a plate and one or two corpi containing drawers. The conditions for the plate are a surface rectangle of approximate geometry (parallelogram or trapezoid) with a maximum tilt and a minimum slant. The conditions for a chest with drawers are a surface of frontal face, containing a nested rectangle. The plate surface and the chest surface should form a foldable surface. There is another characteristic feature grouping that is unique to some desks: parallel aligned 3D rectangles. They occur if the desk has two corpi.

The remaining category representations can be similarly expressed as the two objects described above. With such descriptions however, the *categories were not distinct enough*. There was still overlap between the descriptions, either because they are structurally similar or due to accidental feature grouping, which is often caused by category redundant structure - or what we have specifically termed part redundancy. We therefore had to specify the category representations more

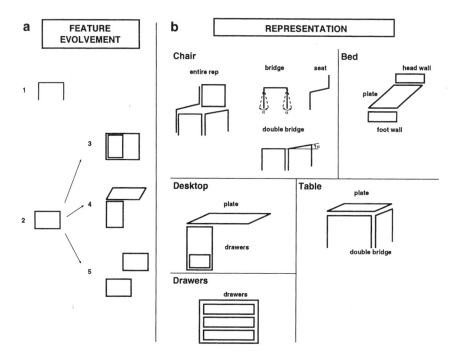

Figure 26: A representation with rectangles (surfaces) and character-istic 3-line polygons. a. Extracted features. b. Category representa-tions with the features shown in a.

elaborately, for example in terms of their geometrical proportions. We describe one example.

The chair structure, as described in the previous paragraph is de-tected in all chairs, in three beds (numbers 1, 3 and 6) and one desk (number 7). The accidental chair in desk number 7 is found in the double bridge of the chest and the drawer rectangle just above. To further refine the chair structure, we introduce a geometrical con-straint on the bridge feature: it is only accepted as such if the sum of the two leg lengths is larger than the length of the middle line. This expresses that the legs of a chair are generally longer in proportion to the seat length than for a bed. With that condition, all the chairs are uniquely distinguished from all other objects, including the accidental chair detection in the desk.

Conclusion It was possible to get the categories distinguished but it required quite a number of structural relations to form a category representation. Our expectation was however to find representations that were easier to form and easier distinguishable from each other.

Despite the use of high-level features, there were still a lot of accidental detections in other categories. In other words, there was still some sort of *representational overlap* between categories. Our intuition was, that there was an important feature of the evolvement and representations missing, namely space (or region). For example, if one knew the immediate surround of a feature, like the surround of a 3-line polygon or a surface, then one gains much more category-specific information. The exploitation of such context is what we explored in our second simulation study.

5.2   A Representation with Polygons and their Context

Simulations   The features we evolve in this simulation are the same in terms of contour geometry as the ones before, but each feature will contain information about its immediate context. This proceeds as follows. After L feature extraction, the structure 'inside' and 'outside' of each L feature is determined (figure 27a, L feature analysis). Looking at the 'emptiness' of these two areas, an L feature is divided into an In, Out or Lone feature. *In* features have one or more lines in their outside but no line in their inside area. *Out* features have one or more lines inside but no line outside. *Lone* features have no lines around their corner, neither in their inside nor in their outside. Lone features are simultaneously In and Out features. As implied with this classification, we are not so much interested in the exact structure around a L feature, but mainly in finding its empty side. To progress toward complex, global features, we start connecting neighboring L features to 3-line polygons. The 3-line polygons, with their classified two corners, contain already an enormously valuable source of structural information that we will exploit for object description. Out of a possible set of 6 U-like and 6 Z-like features only a fraction makes sense in our typical environment (Rasche, 2002b). Some of them are shown in figure 27a, numbers 1-4.

This type of context analysis leads naturally to a description as made of surfaces and silhouette features. Examples of such silhouette features for a chair are: The seat feature is generally a Z-like 3-line polygon made of one In feature and one Out feature; The back-rest is typically a U-like 3-line polygon made of two Out features; The bridge is a U-like 3-line polygon made of two In features. Many of such silhouette features are highly characteristic to a category.

The context of many surfaces is also highly characteristic. Both, the silhouette features as well as the surface features (with context), are distinct enough that it generally suffices to describe a category just by a list of such features, without setting them into structural relation with each other! The exploitation of contextual information vastly reduced the representational overlap between categories and

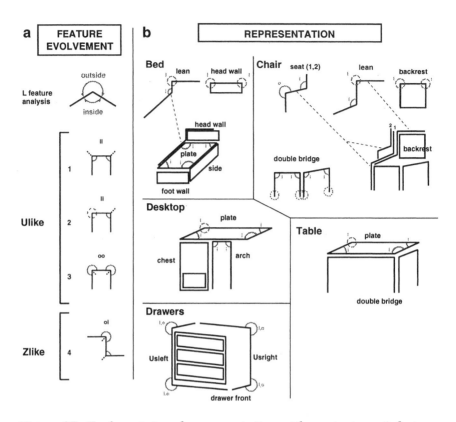

Figure 27: Evolvement and representation with context. a. L feature analysis: its inside and outside. 1-3: some of the useful Ulike features. 4: One useful Zlike feature. b. The context of features is only selectively shown.

many of the geometrical conditions, that we used before - to get the categories laboriously distinguished -, could be dropped.

Conclusion If a representation were a collection of such silhouette and surface features, then *representations would be over-determined*, because many contour pieces of an object are used for surface and silhouette features simultaneously. Consequently, only a fraction of features is necessary to categorize the object. This is not useless redundancy, but advantageous, if not even necessary whenever only a fraction of the set of features is available. This frequently occurs:

1) In real-world scenes in which part of the contours are often of low contrast or ambiguous due to noise and thus not easily detectable, see for example (Canny, 1986) or chapter 6.

2) When we see a novel category instance that has partially new

features. The remaining, familiar features would trigger the category process and the new features could be learned immediately.

3) When objects are in their real-world background, the silhouette contours can be interrupted by contours from objects in the background, e.g. a floor contour. In such situations, surface features are more decisive in the categorization process.

4) Or when the object is in sharp contrast with its background. This may occur when we enter a dark room in which we are only able to see the outline of objects. In this case, silhouette features are more stringent in determining the category type.

This over-representation could also explain, why humans never make categorization mistakes if the object is seen from a canonical viewpoint.

Flexible Recognition Evolvement. The list of items in the previous paragraph implied something regarding recognition evolvement: because there are so many different situations, in which the exact same object can display a different subset of its features, the recognition process has to start with a given feature subset and still yield the same categorization result. Recognition evolvement may thus not progress along a fixed path. A prominent debate relating to this issue, is the discussion whether local or global structure is interpreted first ((Navon, 1977; Palmer, 1999), see also chapter 3). If local structure was interpreted first, then the furniture objects had to be interpreted by their local surfaces and gradually integrated to the global object structure. If global structure was to be interpreted first, then one had to start with silhouette features and work toward local surfaces. Given the previously sketched situations, neither recognition evolvement is preferred: its starting point depends on the situation and its resulting displayed features. Thus, recognition evolvement may be a highly flexible process.

5.3 Recapitulation

The goal was to find a set of descriptors and their loose arrangement that would enable swift categorization. Such representations are supposed to represent the loose perceptual representations envisioned in figure 2 (left side), and specified in chapter 2 (section 2.5). In a first attempt we described the objects by surfaces as well as polygons that are characteristic to a category (Rasche, 2002a). Despite the use of such complex features, there was a considerable overlap in representations that could only be avoided by introducing structural conditions regarding the proportions of a category. From this first attempt we felt that an important, missing feature was the immediate context (or region or 2D space) around features. In a second attempt, we therefore started with a context analysis of each L feature that led to a description with surfaces and silhouettes (Rasche, 2002b). The inclusion of the silhouette descriptors vastly improved the recognition process. In an evolvement including context it is straightforward to figure out which local contour belongs to which other one: context (or space) binds contours. The features, with their attached 'context information', were consequently very distinct from other categories and category representations could be formed using feature lists only. The conclusion that space is essential for evolvement and representation is the foremost insight of this simulation study, and of the work described in this book. And by space is meant *any space* (or 2D region). Not only space engulfed by the contours of a surface, but also space around and between surfaces or parts. That space can be described as silhouette regions. To summarize that in one single phrase: *space binds contours and leads to distinct representations.*

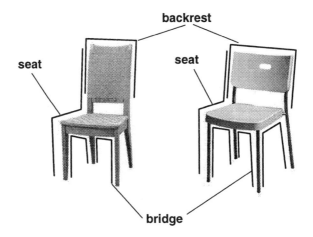

Figure 28: Two chairs and some of their common silhouette regions. The open polygons (three sequentially connected lines) outline some silhouette regions (seat, back-rest and bridge feature), which are largely independent of the structural variability found in the category chair.

With this idea of using space for representation, in particular the silhouette space, let us look at the two chairs in figure 28. Both show quite some part-shape variability, some part-alignment variability and a little bit of part redundancy. But what they share most are the regions, some of which are outlined by the open polygons (three sequential lines). These silhouette features may not always be present in a visual scene, meaning when the object is found in its typical scene context: for example a background contour, like one formed by the floor and the wall, may intersect these silhouette features. We regard this background structure as part of the structural variability aspect, with which the recognition process has to deal. A substantial part of the remainder of this book can be roughly outlined as a quest for a neuromorphic network encoding space, in particular chapter 7 and 9.

Much more can be explored with such line drawings regarding the nature of loose category representation and the nature of an efficient evolvement, but we will now move toward gray-scale images in the next chapter. The primary issue, in particular in chapter 7, will then be, how to efficiently encode space with a neuromorphic network.

6   Retina Circuits Signaling and Propagating Contours

6.1   The Input: a Luminance Landscape

The retina absorbs the luminance distribution of our visual environment and transforms it into a signal useful for the analysis in higher areas in cortex. Before we worry about how this transformation is carried out in neural terms, it is sensible to understand the characteristics of the input and too see how computer science approaches have dealt with it.

If one plotted the luminance distribution (or gray-scale image) as a 3D plot with the third dimension being luminance, then one could see a profile reminiscent of a landscape. Fu has nicely expressed this view by describing it as a *geographic structure* made of plains, slopes, ridges, valleys, shoulders and feet (Lee and Fu, 1981). To extract contours and segments of the image, they created algorithms to detect each of these geographic features. Most other contour-detection approaches have not adopted such a specific viewpoint, but have merely extracted 'general' edges. Each of these algorithms has its advantages and drawbacks. They roughly work in two steps. Firstly, the gradient at each image point (pixel) is determined by looking at its immediate neighborhood, e.g. a 3x3 array with the center pixel taken as the investigated image point. Secondly, neighboring gradients with similar orientations are connected to form contours. The output of such algorithms is thus a list of contours, which can be readily used to filter geometrical features like we did in chapter 5.

The visual system performs this contour extraction a little bit differently. The retina does signal the contours somehow, but the contours stay in the 2D visual field - and are not extracted as lists. This transformation from the luminance landscape to a contour signal is discussed next.

6.2   Spatial Analysis in the Real Retina

The retina consists of roughly 5 neuron classes: photoreceptors, bipolar cells, ganglion cells, horizontal cells and amacrine cells (Dowling, 1987). The first four cells generate only analog potentials: they are non-spiking. The ganglion cells are the only spiking cells, with axons projecting to the thalamus. There exists a so-called direct pathway from the photoreceptors via the bipolar cells to the ganglion cells, whose transformation can be summarized as follows: the luminance (gray-scale) value hitting the photoreceptor determines its analog potential, which is linearly transformed into a spike frequency in the ganglion cell (see figure 29a, number 1 and 2). The analysis of this transform has been carried out with small spotlight stimulations and

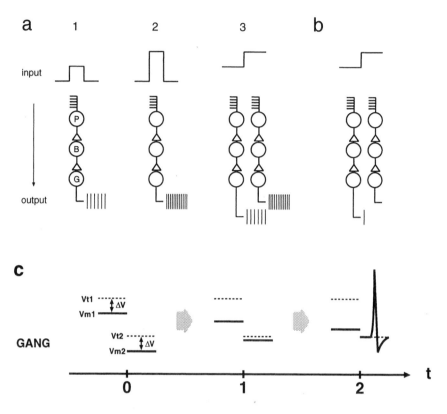

Figure 29: Retinal analog-to-spike transformation. a. Frequency (rate-coded) output. 3 cases: 1) low intensity input, low frequency output; 2) high intensity input, high frequency output; 3) edge input, two frequencies as output. P: photoreceptor; B: bipolar cell; G: ganglion cell. b. Our transformation: an edge triggers a single spike. c. Dynamics of our transformation for two neighboring 'ganglion' cells (time steps 1 and 2). In a fast process (t=0), the edge determines their membrane potentials Vm1 and Vm2, and their spiking thresholds Vt1 and Vt2, which are set above the membrane potential with some fixed voltage offset, ΔV. In a subsequent, slower process, the charge will spread (t=1, t=2) and Vm1 and Vm2 will even out (spiking thresholds stay fixed), and cause the neighboring cell to fire, thus signaling the edge.

has led to the picture that the spatial analysis in the retina occurs point-wise, and that a frequency code is at work. It is however not exactly clear, what happens when an edge is presented (see figure 29a, number 3). If the edge was transmitted as two frequencies, a low and a high frequency respectively, then cortical cells would have to disentangle the frequencies to detect an edge (or contour piece). Because the specific transformation is actually not clear in detail, we here suggest two alternate transformations. One exploits the possibility that spiking thresholds are adjustable and that leads to detection of edges. We call this now the method of adjustable thresholds; the other transformation employs a latency code to separate edges in time. We call this now the method of latencies. In addition, both transformations make use of the idea of wave propagation.

6.2.1   Method of Adjustable Thresholds

Let us assume - as we did in the introduction already -, that the retina signals contours, specifically that it responds to large luminance edges (Rasche, 2004). Furthermore, the output of the neuron should only be a single spike because that can suffice for shape analysis. The transformation process can then be described as: we need a process that generates a single spike in response to a large contrast edge (figure 29b).

We imagine that this may occur with a combination of two separate processes, a fast one and a slow one (figure 29c). In the *fast process*, a receptor potential determines the initial membrane potential and the *adjustable spiking threshold* of its successor (spiking) neuron, which we call ganglion cell now. For reason of simplicity, the second layer (bipolar and horizontal cells) is omitted. The adjustable spiking threshold of the ganglion cells is set above the initial potential with a fixed offset. This fast process is not explicitly shown in figure 29c but merely its 'end result', which is at t=0 of the time axis of the slow process. In the *slow process*, the charge spreads laterally through the network of connected ganglion cells (time steps 1 and 2 in figure 29c). The spiking thresholds stay fixed during this slow process. The charge of a high potential ganglion cell will spread towards its neighboring ganglion cell with a lower potential as well as a lower spiking threshold, and cause it to fire (time step 2).

The motivation for the fast process is that receptors directly determine the membrane potential in their successive ganglion cells, whereas they indirectly determine their adjustable spiking threshold through an extra-cellular process. For example it has been shown for various brain cell cultures, including retinal preparations, that calcium waves can spread quickly through the gap-junctions of the glia network (Charles, 1998). These calcium waves can alter the extra-cellular calcium concentration rapidly and substantially, and could therefore have a significant effect on the electrical behavior of neurons within short time. The charge propagation is motivated by the fact that there exist traveling waves in the retina (Jacobs and Werblin, 1998).

6.2.2 Method of Latencies

The second transformation that we suggest uses the idea that the input magnitude determines the latency of the first spike, a transformation called time-to-first spike by Thorpe (Thorpe, 1990), or generally called a latency code (section 3.3, see specifically figure 13). To implement this, one would continuously feed the neuron with a specific intensity value - it charges up the neuron at a rate determined by this value: hence, high pixel values will trigger early firing, low pixel values will trigger late firing. That will result in signaling bright areas first, followed by signaling of darker areas. Hence, contours are separated in time.

Before demonstrating the operation of the above processes on grayscale images, we firstly introduce the idea of charge propagation in a two-dimensional map and subsequently introduce the above contour detection process.

6.3 The Propagation Map

The propagation map is a two-dimensional array of connected neuronal units (figure 30). Each neuron, depicted as circle, is connected with its eight neighbors via a 'horizontal' resistance. The neuron model is a perfect integrate-and-fire unit (chapter 3, section 3.7), but the exact neuron model does not matter: a different model would just require an adjustment of certain parameters. The map is in some sense the compartmental modeling approach used for approximating dendrites, laid out in two dimensions, with each compartment having a spiking unit.

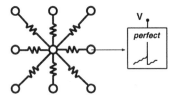

Figure 30: Connectivity of the propagation map. A circle represents a neuron, modeled as an integrate-and-fire type.

If the activity level of a neuron is raised above the spiking threshold of the map, then the generated spike will contribute significantly to the

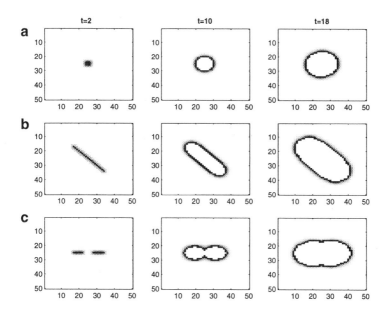

Figure 31: Behavior of the propagation map in response to simple stimuli (Matlab simulation). a. Block source. b. Diagonal line. c. Horizontal line with gap. The snapshots are taken at times t=2, t=10 and t=18, with time equal the number of time steps. Grey: subthreshold voltages. Black: Spikes.

activity level of its neighbors and cause them to fire, thus triggering a traveling spike wave. In order to avoid the bounce back of a wave, the neuronal unit requires a short refractory period. Figure 31 illustrates this propagation process for three different stimuli. In figure 31a the stimulus is a point source: four neurons are activated by raising their activity threshold above the spiking threshold. The spikes are shown in black, the gray values ahead of the spike front represent subthreshold activity. The traveling wave is an annular outward growing wave. For a straight line, the traveling wave is of oval shape (figure 31b).

The propagation map has a characteristic which is extremely useful for encoding space: when the two traveling waves meet, they merge. This is shown in figure 31c for two short lines triggering traveling waves, which eventually merge into a single one. Thus, *contour propagation seals gaps*.

When a luminance landscape is placed into such a map - that does not raise the activity level above the spiking threshold -, then a smoothening process takes place by means of the lateral connections: the landscape starts to even out, but in particularly fast for 'noisy' values that pop out of an area or a local plateau. This subthreshold propagation does not play such a big role for the method of adjustable thresholds, but somewhat more for the method of latencies.

The exact propagation characteristics can be modulated in many different ways (Rasche, 2005a). In the above example, the traveling wave has a width of a single spike, but can be tuned to show a broader width by changing parameter values of the map or by increasing the radius of the local connections. Likewise, the waves can be made faster or slower by changing parameters, something we will exploit to perform motion detection (chapter 8).

6.4   Signaling Contours in Gray-Scale Images

6.4.1   Method of Adjustable Thresholds

We now return to the idea of signaling contours within a network of connected ganglion cells. We take the same map as discussed in the previous section and add the mechanism discussed in figure 29c. We do not emulate the fast process explicitly, but take the luminance distribution of an image directly as the membrane potential of the entire map. A fixed offset value is added to each neuron, which represents the adjustable spiking threshold. Then, in the slow process, the charge propagates and will trigger spikes at steep luminance edges (contours). And once such a contour front is signaled as a line of spikes, it will start traveling across the map. Figure 32 shows this for two objects. After the first two simulation steps (t=1 and t=2), the high contrast contours are already signaled. After further propagation, lower contrast contours are detected as well. As a comparison, the contours found by a popular computer vision algorithm are shown (Canny, 1986).

It should be pointed out that the contours are signaled independently of the absolute luminance level. In a contour profile (of our visual environment) the contrasts can be at any level and the level itself often varies along a contour. The mechanism of figure 29c automatically takes care of that. Thus, there is only one parameter in this network, the offset for the adjustable spiking thresholds. An increase or decrease of the offset value will cause the network to detect higher or lower contrast contours respectively (see figure 33). This offset parameter roughly corresponds to the threshold values used in the Canny algorithm.

6.4.2   Method of Latencies

If one used this method without the laterally connected network (without the propagation network), then noisy pixels are signaled ahead of the area they are placed in. The charge propagation counteracts that a bit, by smoothening the areas slightly and thus leveling out noisy pixels. Figure 34 shows preliminary results with this method.

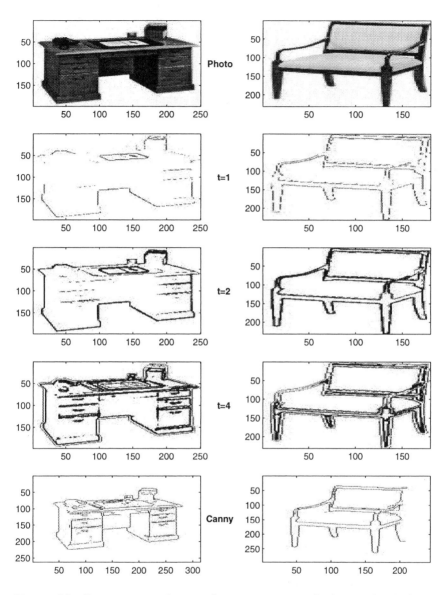

Figure 32: Contour signaling and propagation with the method of adjustable spiking threshold. Top row: photos. 2nd, 3rd and 4th row: Contour propagation after 1, 2 and 4 time steps, respectively. Black lines and dots represent spikes. Bottom row: Contours obtained from the Canny algorithm (finest scale). From Rasche 2004.

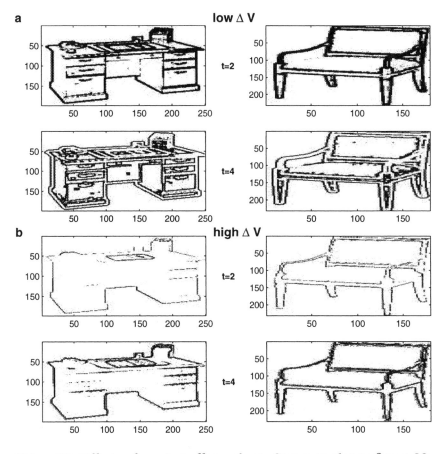

Figure 33: Effects of varying offset values. Compare also to figure 32. a. low ΔV: more low-contrast contours are signaled. b. high ΔV: only high-contrast contours are detected. Time steps 2 and 4 are shown only. From Rasche 2004.

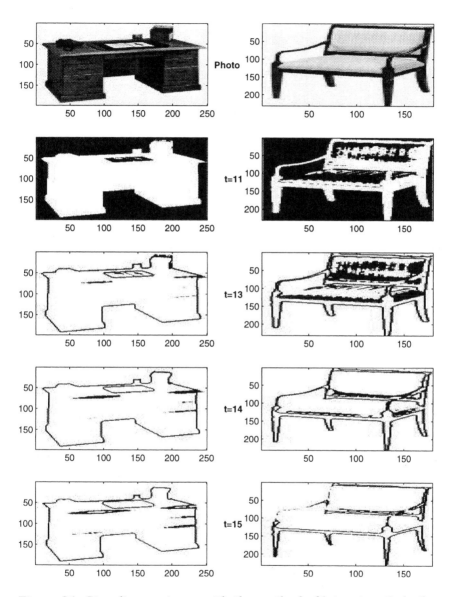

Figure 34: Signaling contours with the method of latencies. Only the spiking output is shown - starting at t=11. Bright areas are signaled first.

At t=11, the first spikes are signaled representing the silhouette areas. No contours per se are signaled within that time slice, but after this the border spikes start to propagate across the lower-intensity (darker) areas (t=13 and later). Darker areas are signaled later, like the lines dividing the drawers. In this specific simulation the signaling of bright areas starts late (at t=11), and causes propagation across many dark areas before those have been signaled. Other contours can be signaled by adjusting the parameter values accordingly, but that needs to be further elaborated. This specific transformation does not offer any contour propagation across the bright areas, but could be achieved by feeding the spikes of this transformation into a subsequent propagation map.

6.4.3 Discussion

The method of adjustable spiking thresholds signals contours relatively immediately as compared to the method of latencies and propagates them across its map. It therefore represents a compact way to obtain contours and their propagation simultaneously. The method of latencies is somewhat more intricate: the contours are stretched out in time and the retinal network does provide only partial propagation. But it offers the following advantages. Firstly, it may be easier to implement into analog hardware than the method of adjustable thresholds, whose fast process is not explicitly simulated. Secondly, if one looked at a neuron 'patiently', then one would observe a firing rate which actually reflects the intensity of the pixel. This latter point has already been suggested by Thorpe: the latency information would provide fast computation, the rate information would provide slower computation. Specifically applied to the purpose of categorization, the latencies provide enough for generating the contour image and subsequent perceptual categorization, but for determining other aspects of visual information, which may happen on a slower time scale, a rate code may serve well.

6.5 Recapitulation

We have started this chapter by reviewing Fu's description of the luminance profile as a geographical map. We have not specifically addressed the diversity of this profile, but one should keep it in mind, if one plans to extend this approach into low-resolution gray-scale images in which structure can be very subtly embedded.

We have proposed two methods for contour extraction. The method of adjustable thresholds is a process that detects steep gradients (contours) in a luminance profile. Initially, very high-contrast contours are signaled, followed by signaling lower-contrast contours. Due to the propagation process, gaps in the contour image are filled by the expanding and merging propagation process. To put it into a succinct phrase: *contour propagation seals gaps.* In the method of latencies, the edges are separated in time.

The output of the retina looks very similar to the output of an edge-detection algorithm used in computer vision. Still, the obtained contours are fragmented - as they are with any method performing contour detection. In computer vision, much effort has been directed toward obtaining a more complete contour image, often combined with efforts that are called image segmentation (Palmer, 1999). Yet, the fragmented contour image already provides an enormous wealth of structural information: it delineates many regions as we have determined them in our line drawing studies. The output therefore suffices for the perceptual categorization we aim at (figure 2, left side). The loose representations that we search for, have to be able to cope with part-shape variability, part-alignment variability and part redundancy anyway (section 2.1): It therefore does not matter, whether there is one or the other contour missing after the contour extraction process. This lack of contour pieces is likely to the smaller problem in the construction of a visual system - if it is one at all -, than the challenge to deal with structural variability.

The simulations presented so far have been merely software simulations. How either retina model can be translated into analog VLSI remains to be worked out. A starting point would be to develop the propagation map using maybe the propagation tools for neuromorphic dendrites (section 4.3); in the next step, one would insert the contour detection mechanism.

7 The Symmetric-Axis Transform

From our line drawing studies in chapter 5 we have gained the intuition that encoding space is valuable for evolvement and representation. The empty space between contours helps to determine which local contours belong together and that leads to a description by surface and silhouette regions. The goal in this chapter is to find a neural way to encode space. We thereby seek inspiration from the two mechanisms that we have discussed in chapter 3. One mechanism is Blum's symmetric-axis transform, the other one is the Fourier-like filter (or channel) approach. In this chapter we focus on Blum's symmetric-axis transform because it operates more specifically and because it returns a shape description useful for high-level representations. We show how to simulate this transform with a neuronal architecture made of integrate-and-fire neurons.

7.1 The Transform

The symmetric-axis transform (SAT) is a process in which the contours of a shape are propagated across an excitable map and wherever two propagating contours meet at a given point in time, they leave a symmetric point (sym-point). The loci of sym-points form the symmetric axis (sym-ax), which completely defines the contour geometry of the shape and its including space (Blum, 1973). Figure 35a shows the SAT for a rectangle at three stages: at the beginning (0), in the middle (mid) and at the end (final). To understand the usefulness of the sym-axes, their completed evolvement is plotted in a 3D space with dimensions x, y and time (figure 35b). In this 3D space, the sym-ax of the rectangle forms a roof shape, made of five (straight) *trajectory pieces*. Four of them point from the rectangle's corner toward the corresponding ends of the middle line, which in turn is evolved last and at once. For a circle, the sym-ax is a single point, which occurs at a time depending on the size of the radius (not shown). For an ellipse it is a sym-ax made of two increasing trajectories running toward each other and meeting in the center of the ellipse (figure 35b, right side). All these trajectories are easily graspable and therefore convenient for higher level representations.

The SAT readily encodes the inside of a shape. It is therefore useful to extract many of the features we have developed in our line-drawing study (chapter 5). It does not, however, encode the outside of an L feature or shape. For instance, the rectangle's silhouette is not directly encoded. In a later chapter, we will offer possible solutions to this lack of 'outside' encoding (section 9.1).

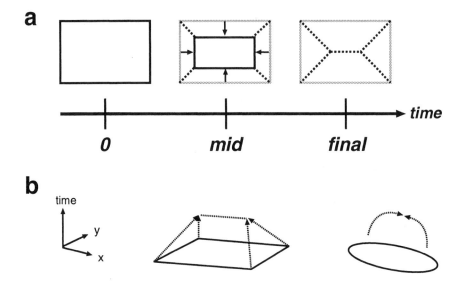

Figure 35: The symmetric-axis transform (SAT). a. Evolvement of the symmetric axis (sym-ax) for a rectangle. b. 3D plot of the sym-ax trajectories for a rectangle and an ellipse.

7.2   Architecture

Outline   Blum envisioned that the SAT can be implemented by an excitable map with the following properties for each point: '1) it has value either one or zero (all-or-none excitation); 2) it may be excited by an adjacent point or by an external source; 3) an excited point cannot be re-excited for some interval of time (refractory property)'. This list of properties suggests to use a single map of I&F neurons. Yet, there is a difficulty with using only a single map: it is tedious to distinguish whether two wave fronts coincide or whether there is only a single wave traveling across a given point (neuron) in the map. We imagine that this is easiest solved by a selective mechanism: it is able to clearly sense when two waves clash and to ignore when a single wave travels across the map. We therefore decided to separate the process into *three 'layers'* (figure 36): a propagation map, a set of wave-detecting orientation columns and a coincidence detector map. The propagation map (PM) is an excitable membrane that actively propagates contours like the propagation map presented in chapter 6 (section 6.3). To sense whether two distinct waves are about to collide, we were inspired by the orientation column architecture of the primary visual cortex. As reviewed in chapter 3, V1 neurons are believed to perform edge detection or spatial filtering of oriented structure. But it has also been suggested that they may be involved in region encoding (Palmer, 1999). And that is how we exploit the orientation column (OC) structure: V1 cells detect when a wave of certain orientation propagates through their receptive field. Put shortly, *'V1 cells' act as wave-detectors*. The top map, the symmetric-axis map (SAM), signals sym-points by detecting when two distinct waves are close. It does that by correlating the activity of the V1 cells.

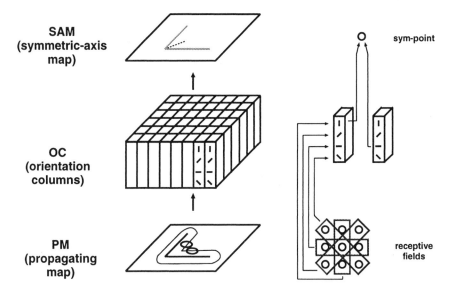

Figure 36: SAT architecture. Processing occurs from bottom to top. The propagation map (PM) is an excitable membrane that allows contours to travel across it. The orientation columns (OC) contain an array of neurons that sense the orientation of waves in the propagating map and fire a single spike when a wave of preferred orientation travels through their receptive field (RF indicated by circles). The symmetric-axis map (SAM) evolves the sym-points by sensing when two proximal orientation-selective neurons of different orientation fire.

Implementation   An implementation of the propagation map has already been discussed in section 6.3 (see also (Rasche, 2005b)). The output of this propagation map is then sent to the OC structure. An OC neuron fires a single spike, whenever a wave piece of preferred orientation travels through their receptive field. To obtain this selectivity, the neurons need to be correspondingly wired (1) and possess dynamics that filter the 'coincidence' of its input (2).

1) Wiring: Each neuron receives input from three linearly aligned neurons of the propagating map (figure 36, 'receptive fields'). For

simplicity, we employ only 12 orientations (only 4 orientations are shown in figure 36). The wiring alone does not achieve the preferred orientation-selectivity because a wave of any orientation - propagating through the receptive field - will stimulate the receiving neuron. The neuron is therefore modeled as a coincidence detector:

2) Coincidence: A neuron is modeled as a leaky integrate-and-fire type, which possesses a constant leakage, continuously draining from the neuron's activity level (section 3.7). Due to this continuous leakage, activity input to the neuron is only transient. In order to efficiently integrate, its synaptic input has to occur coincidentally. The neuronal dynamics are tuned such that the neuron will only fire when its input neurons fire simultaneously, which only happens when a wave of appropriate orientation travels through their receptive field. If the wave has non-preferred orientation, then the synaptic input occurs sequentially, which is insufficient for integration to the spiking threshold.

The symmetric-axis map is a two-dimensional sheet of neurons, also possessing specific input wiring and specific neuronal tuning. Each neuron receives input from proximal orientation-column neurons, whereby the exact pairing of orientation-selective neurons does not matter, because a sym-point is supposed to signal the coincidence of any two waves that are about to collide. An exception are pairs of orientations that are of similar angle and sequentially aligned, because such a pairing would also signal if a contour was curved. Because of this dilemma and because we use merely 12 orientation angles, only L features of small angle are encoded (smaller than 110 degrees approximately).

The neurons of the symmetric-axis map are also modeled as leaky integrate-and-fire neurons, because they need to detect the coincidence of its input, analogous as the orientation-selective cells do. The neuron dynamics are adjusted such that the symmetric-axis neuron fires a single spike, when its two input neurons fire simultaneously.

7.3 Performance

Space as a Structurally Robust Feature Figure 37 shows how the implementation performs for a set of L features and should demonstrate the usefulness of encoding space. Although the various L features have structural differences, their space between the two lines is similar for all of them. The sym-ax for the dashed and curved L starts with a delay compared to the sym-ax of the solid L, but their direction and endpoints are the same, indicating thus the similar space but also the similar contour geometry of the feature. In a pure contour description approach in contrast, the dashed L had to be described by integrating the individual line pieces, which is an expensive undertaking. Using the neuromorphic SAT, the space between the legs

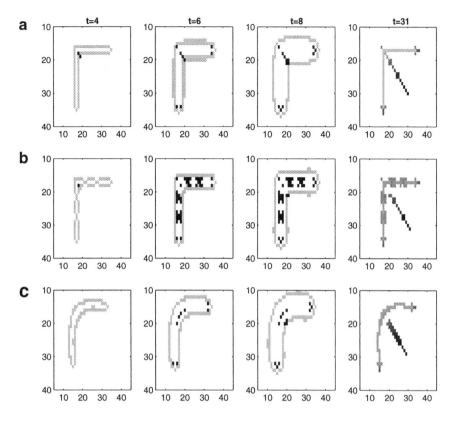

Figure 37: Sym-ax evolvement of varying L features. a. solid L. b. dashed L. c. curved L. The snapshots are taken at times t=4, t=6, t=8 and t=31, with time being the number of time steps. For t=4,6,8: grey dots or lines represents spiking activity in the PM; black represents sym-points as evolved in the SAM (no subthreshold activity shown in this figure). For t=31: lightest gray: original L shape; grey to black: sym-ax evolved in time.

is exploited and provides a large amount of robustness to incomplete contours and structural variability, in particular part-shape variability. At the core of this robustness is the sealing property of the PM as discussed in section 6.3 (see figure 31c).

Gray-scale Images The performance of the SAT in gray-scale images is illustrated in figure 38. The top row shows the objects, a chair and a desk. The middle row shows the contours that have been extracted using a computer vision algorithm (Canny, 1986). It would of course make sense to employ the retina of chapter 6, but we have not devel-

oped a synthesis between these two systems yet: it required modifica-
tion of the coincidence-detecting OC cells. The bottom row shows how
the regions are encoded. These encoded regions readily correspond to
features as developed in our line drawing simulations (chapter 5). For
example the regions between the chair's legs are encoded, which we
termed bridge features in chapter 5.

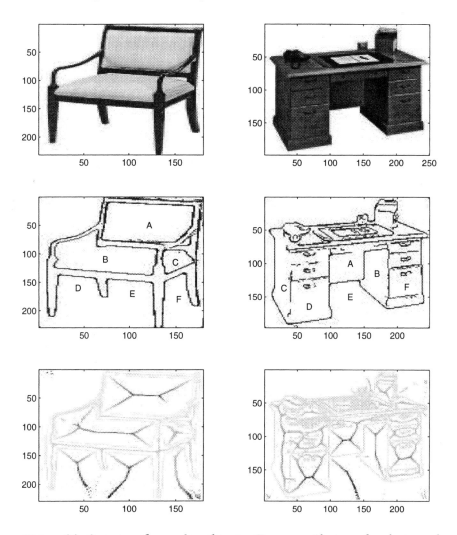

Figure 38: Sym-ax of complex objects. Top row: Photos of a chair and
a desktop. Middle row: Contours extracted with the Canny algorithm
at finest scale. Bottom row: Completed SAT. The major spaces (regions) have been encoded. The object contours are shown in lightest
gray as in the figure before.

The enormous advantage of the SAT lies in its robustness - as we have already tried to express with figure 37. For example, the chair's legs could possess ornament; the back-rest could be slightly differently shaped; the desk's drawers could be of rectangular shape with rounded corners, and so on. Despite this part-shape variability, the resulting sym-axes are structurally very similar, they are closely spaced in the 3D space mentioned in figure 35.

Albeit this benefit of robustness, the SAT can not capture certain silhouette features as we have already mentioned before. To express it with the feature language used in chapter 5, it is Out and Lone features that are not encoded: For instance, the space around the back-rest or the space around the desktop's plate is not encoded. One can find these features by analysing the spatial relations of the sym-ax pattern: for example, if the starting point of a L vector has no other starting points (of neighboring L vectors) in its surround, then the starting point of that vector represents a Lone feature. If one analyzed the sym-ax trajectories of an object using a (computer vision) list-approach like we did in chapter 5, then one could readily determine the context of the trajectories. If one analyzed the sym-ax trajectories with neural networks, then one had to find either a method to determine the context of the trajectories or to find a substrate that recognizes the 'outward' propagating contours. We discuss some possibilities in chapter 9.

Another short-coming of the existent architecture is that 'speckled' noise or sometimes fine texture in the middle of regions may trigger traveling waves that prohibit the proper sym-ax encoding of the region. We can think of a number of possibilities to deal with that. One would be to introduce some contour grouping mechanisms like Marr did to filter out the long-connecting contours (Marr, 1982). Such contour grouping mechanisms are akin to the Gestaltist's ideas of perceptual grouping (Bruce et al., 2003). Another possibility would be to modify the wave propagation map and to let contours propagate depending on their qualities: high-contrast contours would trigger large waves, low-contrast contours would trigger small waves. Some of the speckled noise tends to be of low contrast and would thus be swallowed by large waves from high-contrast contours. Contour propagation could be made less active in the sense that contours would only travel a certain distance, depending on their length or some sort of energy. That would also favor long contours, and certain noise and texture would be ignored. Such 'segmentation' efforts however, should not result in the pursuit of immaculate segmentation, which we regard as impossible: the contour image obtained from our retina (or a computer vision algorithm) is far sufficient for much of the perceptual categorization we pursue in this book.

7.4 SAT Variants

Blum's SAT has already been implemented by computer vision scientists using a variety of computer algorithms (see (Ogniewicz and Kubler, 1995) for a review). These implementations have been applied to problems like shape identification (Pizer et al., 1987; Leymarie and Levine, 1992), letter recognition (Kegl and Krzyak, 2002) and medical image analysis (Dill and Levine, 1987) to mention only a few examples. These approaches are partially inspired by a variant of the original SAT, which aimed at reconstructing 3D bodies from images (Blum and Nagel, 1978). Many of these algorithms are concerned with an exact reconstruction of the shape and often require closed and complete contours. Some of them iterate until the exact sym-ax - reflecting the detailed shape - has been evolved. The purpose of our SAT is different. It transforms the major regions into coarse sym-axes which are suitable for forming the abstract category representations we look for. During the transformation, a lot of structural detail is washed out, which can be omitted for the purpose of categorization. The SAT presented here can operate on incomplete contours and is performed in a single sweep, that means without iteration.

Other vision researchers suggested a region-encoding mechanism, that turns space into smaller regions, so-called cores (Burbeck and Pizer, 1995).

7.5 Fast Waves

As mentioned before one can regard the propagation of contours across the propagating map as a traveling wave. Traveling waves have been measured in the nervous system of a number of animals (paragraph 'waves', section 3.4). Some of those reported waves are slow though - propagating at a speed of on the order of millimeters per second (e.g. (Hughes, 1995; Prechtl et al., 1997; Wilson et al., 2001; Shevelev and Tsicalov, 1997)). That is too inert to account for the high processing speed (chapter 1 and 3) and can therefore not be responsible for region encoding. Faster waves were measured by Jacobs and Werblin: a (visual) square triggered inward-propagating waves in the salamander retina, which would collapse after a few hundreds of milliseconds (Jacobs and Werblin, 1998). This maybe just sufficiently fast to account for region encoding . Possibly even faster could be traveling waves in the presence of oscillatory activity: Ermentrout and Kleinfeld demonstrate in a computational study, that coupled oscillators can evoke traveling waves (Ermentrout and Kleinfeld, 2001). They envision that such waves are possibly involved in different neural computations. We can think of some other biophysically plausible alternatives. One alternative is that real neurons could operate close to spiking threshold and act as coincidence detectors, avoiding thus costly (membrane potential) integration time. Studies show that cortical neurons may indeed operate as rapid coincidence detectors (Koenig et al., 1996). A traveling wave 'riding' on such coincidence detecting neurons could therefore be much quicker. Another alternative would be propagation through gap junctions, which transmit electric charge much quicker than chemical synapses and therefore minimize integration time. It was long believed that gap junctions hardly exist in neocortical neurons (Shepard, 1998), yet they are difficult to discover. Two recent studies in rats have indeed found a neocortical network of electrically coupled inhibitory neurons (Gibson et al., 1999; Galarreta and Hestrin, 1999). Another possibility for fast contour propagation are the existence of horizontal connections amongst cortical neurons of primary visual cortex (see (Kovacs, 1996) for a review). Its discoveries have helped loosen the picture that the primary visual cortex is involved in local, patch-wise analysis only. The existence of these far-reaching connections has led some researchers to propose that region processing and perceptual grouping processes may take place in V1 already (reviewed in (Kovacs, 1996; Hess and Field, 1999)). Contours 'spread' along these connections could occur very swiftly due to their far reaching connections. In summary, we see a number of biophysically plausible mechanisms that could provide fast traveling waves or other forms of contour spread that would serve the purpose of rapid encoding of visual space. Such fast wave processes may however be difficult to discover (Glaser and Barch, 1999).

7.6   Recapitulation

The goal in this chapter was to encode space in a neural manner. We have done that using Blum's symmetric-axis transform (SAT), because it is a precise method to turn a region into a trajectory that can be easily used for high-level representations. The transformation was constructed with three separate network structures: 1) a layer performing contour propagation; 2) an orientation column architecture in which 'V1 cells' act as *wave detectors*; 3) a layer that detects sympoints. The transform encodes closed regions, even if their contours are fragmented and incomplete. The sym-axes naturally express many regions as determined in the line-drawing studies (chapter 5). However, the outside of a shape is not captured by the SAT and certain silhouette regions, in particular Out and Lone regions are therefore not encoded. We discuss a possible fixture for the lack of those silhouette regions in chapter 9 (section 9.1).

The SAT swallows some of the part-shape and part-alignment variability by generating sym-axes which represent the region without the exact geometry of the bounding contours. This is a property, which is actually based on the contour propagation map (see section 6.3). Because of this structural robustness, we imagine that the sym-axes can form a significant component of those abstract and loose representations necessary for fast categorization (see figure 39, compare to figure 2).

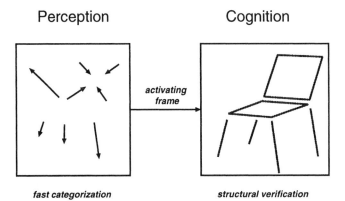

Figure 39: Recognition evolvement with sym-axes. Sym-axes may be a significant component of perceptual category representations. Compare to figure 2 in chapter 1.

A Hybrid Evolvement Process   The simulations presented are only software simulations so far, but are conceptually simple enough to be emulated in the framework of the silicon cortex project (chapter 4). And if it were implemented in analog hardware, the SAT could take place within tens of milliseconds - depending on the speed of the wave propagation process: The transform would be carried out at the same speed as the real visual system performs region encoding processes using traveling waves. This comparison only makes sense if the visual system does indeed use such region encoding. If a retina like the one in chapter 6 was developed, and if a neuromorphic SAT was connected to it, then one had a pure neuromorphic front-end generating sym-axes. How the sym-axes maybe read out we discuss in the following two chapters. Alternatively, one may already envisage a *hybrid categorization system*, in which in a first stage this neuromorphic front-end would be used to swiftly generate the sym-axes, and then, in a second stage, the resulting sym-axes are read out by a computer vision 'back-end'.

8 Motion Detection

The symmetric axes, that we have evolved in the previous chapter, represent a trajectory in a 3D space. For L features, that trajectory is a vector having a certain direction and certain speed, very much like the vector of a motion stimulus. If one attempted to create a neuromorphic substrate that reads out such trajectories, one may as well seek inspiration from models that perform motion analysis and this brings us to the topic of motion detection.

8.1 Models

8.1.1 Computational

The earliest, thorough experimental and computational study of motion detection is the one by Reichardt on the fly's motion system (Reichardt, 1961). Reichardt measured systematically the fly's turning response to a large number of schematic motion stimuli. From the collected response patterns he dervied an elaborate algorithm, that consists of a number of stages performing various types of *filtering and correlating* between inputs. Its very basic gist is shown in figure 40a. The input signals converge to a unit that correlates them: correlation is only successful if the two signals are received in sequence and in preferred order. If the stimulation direction is reversed (non-preferred), then the unit reports no correlation. There is a number of computational studies that is roughly analogous to the Reichardt approach: these studies model the responses of humans to certain motion stimuli and motion illusions (Adelson and Bergen, 1985; Vansanten and Sperling, 1984; Watson and Ahumada, 1985).

Several neuromorphic implementations of the Reichardt principle already exist in analog hardware (e.g. (Tanner and Mead, 1986; Kramer et al., 1995)). They are implementations of the proposed algorithms using analog circuitry performing the necessary types of filtering and correlation. These motion detectors work well if the motion crosses a substantial part of their visual field, meaning their 'receptive field' corresponds to the entire visual field. But one may also intend to develop an implementation in which a motion stimulus is analyzed, when it has only crossed part of the visual field, for example only the left half of the visual field, or speaking in terms of sym-axes, when a trajectory is generated in only part of the visual field. A solution to that may lie in a neural emulation using a retinotopic map.

8.1.2 Biophysical

The computational models mentioned previously have not been specific about an exact biophysical implementation but the proposed algorithms lend themselves toward a neural interpretation. For example

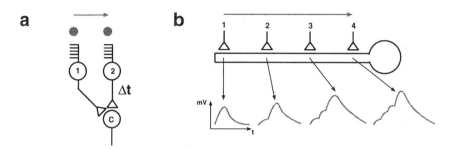

Figure 40: Variants of the delay-and-compare principle. a. Photoreceptors 1 and 2 receive motion input (indicated by two dots). The
elicited signals are correlated in the unit C, whereas the second signal
arrives with a certain delay (Δt). Stimulation in the reverse direction
would not yield a correlation between the signals. b. Schematics of a
dendrite (rectangular shape) and a soma (circle). The four synapses of
the dendrite receive sequential inputs from 1 through 4. Each synaptic stimulation contributes to an increasing wave running towards the
soma. The 'comparison' would take place in the soma.

the correlation unit C (in figure 40a) could be simply a neuron summing or multiplying postsynaptic signals. Because neurons can possibly perform such calculating operations (section 3.7), and because
such operations have been proposed (Barlow and Levick, 1964; Koch
et al., 1983), this may be one way to develop a retinotopic map for
motion detection. But there are also alternative biophysical models:
One on the dendritic level, and one on a 'map' level.

Dendritic Level A model for the dendritic site comes from Rall, who
applied cable theory to the study of dendritic processing (Rall, 1964)
(see section 3.7). He proposed that a dendritic cable can act as a direction sensitive 'wire' (figure 40b): If one stimulates a dendritic cable
sequentially - with stimulations approaching the soma -, then an increasing activity wave runs towards the soma, causing the soma to
spike. If stimulations occured into the opposite (non-preferred) direction, the wave would run into the tip of the dendrite where it has no
effect. This type of *dendritic direction selectivity* has been modeled with
compartmental models and some researchers have interpreted neurophysiological data according to it (Livingstone, 1998). It has also been
implemented into analog hardware ((Elias, 1993; Rasche and Douglas,
2001), see also section 4.3).

Map Level Because of the abundant research performed on a neuronal level, it is tempting to think that the neuron is the sole site

of computation. But motion may also be encoded at the map level, as Glaser and Barch proposed it (Glaser and Barch, 1999). In their model, a motion is encoded by an excitable membrane, that triggers traveling waves. We have mentioned their model in section 3.4 already.

Speed estimation Biophysical models tend to explain direction selectivity only. Less effort has been invested into inventing biophysical models that estimate the speed of object motion. To solve this task one could build on the above mentioned neural mechanism by introducing a range of different synaptic dynamics covering different speeds. Fast dynamics ('short' decay time constants) would be used for detecting high speeds, slow dynamics ('long' decay time constants) would be used for low speeds. If such a scheme is used by the nervous system, then one might possibly find *different synaptic decay times*. Indeed, the nervous system possesses a variety of excitatory and inhibitory synaptic dynamics (Johnston and Wu, 1995). For example, the alpha-amino-3-hydroxy-5-methylisoxazole-4-proprionic acid (AMPA) synapse releases a post-synaptic current (PSC) of short duration (several milliseconds), the N-methyl-D-aspartate (NMDA) synapse releases a current of long-duration (several tens of milliseconds) - the latter however only after the membrane potential has already been elevated to a certain level (see figure 19b, right side; section 3.7). Inhibitory synapses, like the gamma-aminobutyric acid (GABA), also come in variants with different dynamics. The use of such synaptic dynamics may however be too intricate to cover an entire range of speeds. We therefore propose an alternative, which solves the problem on an architectural level.

8.2 Speed Detecting Architectures

One possible solution is to build a pyramid of different neuronal layers made of direction-selective neurons with higher layers receiving only intermittent input from lower layers (see figure 41). At the bottom level would be motion-detecting neurons sensitive to any speed. Their output would feed into multiple layers. Each layer is governed by the same neuronal dynamics, but would receive *intermittent input* from the bottom level: A low level would receive input with small intervals and be sensitive to low speeds; higher levels would receive only sparse input (large intervals) and therefore only detect high speeds. Such an architecture, hereafter referred to as the *speed-pyramid*, would depend to a lesser degree on the need for different neuronal dynamics.

We raise another alternative: speed may be computed with maps of different dynamics (Rasche, 2005c). The idea to detect motion with dynamic maps was introduced by Glaser and Barch (Glaser and Barch, 1999): they simulate an excitable membrane of locally interconnected

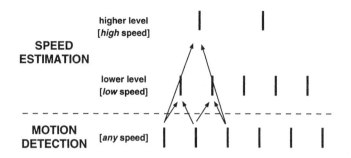

Figure 41: Speed pyramid. The bottom level is sensitive to motion for any speed (MOTION DETECTION). It feeds into layers that estimate speed (SPEED ESTIMATION). Higher levels of the speed pyramid receive only intermittent input (with large intervals) and would thus code for high speeds only.

neurons, which - when stimulated - propagates activity into all directions. Stimulating such a map with a moving input induces characteristic propagation patterns, some of which can explain certain types of motion illusions. Motivated by that study and by our own work on contour propagation using similar excitable maps, we introduce here propagating maps that respond to a motion stimulus of preferred speed.

We basically take the same propagation map as introduced in section 6.3 but tune its dynamics subtly different: the *propagation dynamics are made inert*, such that when the activity raises above the spiking threshold only a single spike is generated. In order to obtain continous spiking in the map, the motion has to stimulate the map continuously at the same speed. Synaptic input placed into such a propagation map will spread to all directions and decay in amplitude similar to the propagation properties of synaptic input in a dendritic cable. Sequential synaptic input along any direction will gradually integrate and eventually reach the spiking threshold and cause spiking. In order to differentiate between different speeds, the dynamics of the map are adjusted correspondingly to filter only a certain speed or range of speeds. An important advantage of such *speed maps* is that they are independent of direction. That is, the tedious placement of direction-selective cables or the spatial arrangement of synapses is unnecessary. Summarized roughly, these speed maps embody Rall's dendritic delay-and-compare scheme but irrespective of motion direction. We have simulated such speed maps and report about them in the next section.

8.3 Simulation

Tuning We firstly try to understand how to tune the parameters of a propagation map to make it behave as a speed map. As mentioned above, the speed-map dynamics that we seek are inert as compared to the dynamics of the propagation map. The propagation map propagates its input *actively* (section 6.3). The speed map, in contrast, is supposed to propagate its input only *passively*, meaning it should decay away - even if input has been raised above the threshold. Thus, the horizontal resistance of the speed map is generally much higher to prevent active propagation. To fine-tune a map to a specific speed, we need to understand the general propagation properties through the resistive network in more detail. The properties are analogous to the propagation properties of a dendritic cable. Changing the axial (horizontal) conductance will primarily change the distance with which a signal spreads: a high value results in far spread, a low value in short spread, or put more formally, a high and low space constant respectively. Changing the axial conductance also affects the decay time constant, to a smaller extent though: a high value causes a faster decay of activity at a given neuronal unit, a low value causes a slower decay. Changing the amount of leakage, L, will primarily modulate the decay time constant: a low value results in a slow decay, a high value results in a fast decay, or a long and short time constant respectively. It will secondarily affect the space constant: a low leakage causes farther spread, a high leakage causes shorter spread.

In order to detect different speeds, the above parameters are adjusted correspondingly. To detect fast speeds, the axial conductance is set to a high value to allow for quick and far spread resulting only in synaptic integration when there is a rapid sequence of EPSP drops. To avoid an integration of low speeds, the leakage is set high. To detect slow speeds, the reverse applies: a low value for the axial conductance avoids the fast run-away of activity and only slowly transmits activity to neighboring units. Additionally, the leakage conductance is set to a small value in order to give slow motion stimulation a chance to integrate.

Simulations Three differently tuned maps were employed, with fast, medium and slow dynamics; each was tested with 10 different speeds (1: slow, 10: fast). Figure 42 shows the response of each map for selected speeds. The stimulating shape was an arrow consisting of 5 dots pointing towards the right side (equivalent to a '>' sign). A plot shows map's spike occurrences with gray-scale values indicating the time of firing (bright and dark dots represent early and late spiking, respectively).

For the fast map, there was only spiking for speed numbers 8 to 10. For the medium map, there was spiking for a larger range of speeds, numbers 3 to 10 (only 3, 5 and 10 shown). For the slow map, the map responded to every speed (only 1, 5 and 10 shown). For each of

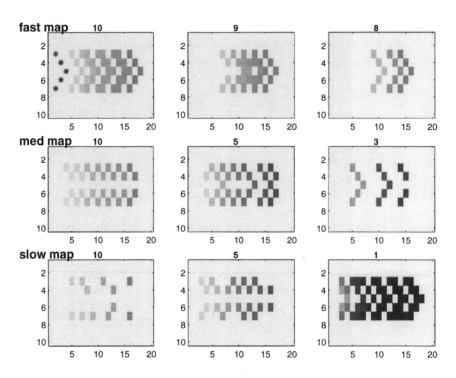

Figure 42: Map spiking in response to a moving arrow shape (shape indicated in upper left subplot with asterisks) for three dynamically different maps (fast, medium and slow) and three selected speeds. All spike occurrences are plotted into the map (size: 10x20), with bright and dark dots representing early and late spiking, respectively (no subthreshold activity is shown). Fast map: response to speed numbers 10, 9 and 8 is shown. Medium map: response to speed numbers 10, 5 and 3 is shown. Slow map: response to speed numbers 10, 5 and 1 is shown.

these maps, there was more spiking for its preferred speed than for other speeds. This is expressed with a spike histogram as shown in figure 43: The total number of spikes for the entire motion stimulus is plotted as a function of each speed number. We call these curves from now on *speed-tuning curves*. For the fast map, the speed-tuning curve starts at speed number 8 and increases with higher speeds. For the medium map, the tuning curve starts at speed number 3 and shows signs of decrease at speed numbers 9 and 10. For the slow map, the tuning curve covers the entire speed range, but declines towards higher speeds. Summarized roughly, for the fast map the tuning curve is sharp, for slower maps, the tuning curve is broad. Thus, the speed

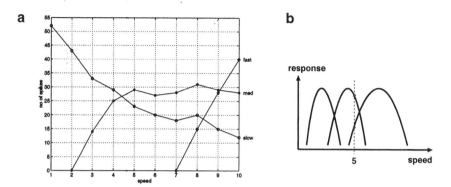

Figure 43: Speed-tuning curves. a. Obtained from simulation for each map (fast, medium, slow) for a dot stimulation (dashed-dotted lines) and an arrow stimulation (solid lines) stimulation. X-axis: total number of occurred spikes for the entire stimulation. Y-axis: speed number. b. Schematic version of a.

maps are only useful for a rough estimate of speed. In order to determine speed accurately, one could possibly use more maps and try to achieve a narrow tuning for each one, which however would likely result in a tuning ordeal. A better solution would be the employment of a speed pyramid (section 8.2), in which one used the same dynamics for each level. In either case, the creation of such a set of maps to cover all speeds is costly in terms of size - thinking in engineering terms. Instead, it is more size efficient and even more accurate to create a set of neurons reading out a small number of tuning curves, see figure 43b. This idea is analogous to the formation of color sensation with only three photoreceptors tuned to three different luminance sensitivities. For example, a speed-detecting neuron would read the ratio of firing rates from two or more speed maps (see speed number '5' in figure 43), and so accurately determine speed, despite the broad tuning-curves.

8.4 Biophysical Plausibility

If the nervous system used such propagation maps for speed estimation, how might the cortical tissue emulate them? And where are they likely to be found?

Regarding the biological emulation, it is firstly the horizontal connection that demands an interpretation, in particular the subthreshold propagation through the horizontal connections. We have already looked for such horizontal activity spread in the section on 'fast waves' (section 7.5): there we were seeking possibilities that would allow for active propagation of contours. Here, we search for a substrate that rather passively propagates activity and that allows for different dynamics selecting different speeds. One possibility for such propagation is the extracellular activity spread through the glia substrate (e.g. (Charles, 1998)). We already suggested this possibility in the retina chapter to motivate the change of spiking thresholds (section 6.2). It was also pointed out by Koch that such extracellular dynamics are not well understood, yet worth to model (Koch, 1999). There is a number of other studies that report about some sort of activity spread through the neural substrate (e.g. (Grinvald et al., 1994; Beggs and Plenz, 2003)). Another possiblity could be activity spread through dendrodendritic synapses. If any of these possibilities could be the basis for passive spread useful for speed detection, then there may also be different dynamics, which would allow for detection of different spreading speeds. Different dynamics maybe caused by different 'packing' density of neurons. Alternatively, evolution may have evolved a speed pyramid, which would not require different dynamics but just sparser connections with higher levels (figure 41).

Regarding the location of such maps, one may readily suggest that they exist in area MT (V5), where many neurons seem to signal exclusively for speed ((Perrone and Thiele, 2001), but see also (Priebe et al., 2003)). But some neurons in primary visual cortex seem also to fire for speed only (see (Nakayama, 1985) for a review). A thorough review of these studies may give hints about whether the brain uses for example a speed pyramid. For instance, we imagine that 'speed-neurons' in lower areas like the primary visual cortex, may fire for lower speeds, whereas higher areas may possess neurons firing for higher speeds.

In some motion detection studies, a reoccuring issue has been the *aperture problem*, which addresses the difficulty to estimate the overall, global direction of object motion from local receptive fields. Local receptive fields, as they exist in primary visual cortex, only sense a subset of the visual field and hence the entire object. An edge wandering through the receptive field may therefore not reflect the global motion direction (reviewed in (Nakayama, 1985)). Furthermore, using only local receptive fields, it is also difficult to estimate the exact speed due to that aperture ambiguity. Attentional networks may solve the problem (e.g. (Nowlan and Sejnowski, 1995)), but here we have proposed a speed-estimation architecture, in which this problem does not really appear, because the architecture does not rely on any local receptive fields. In our architecture, motion direction is not computed at all, but merely motion speed is sensed.

8.5   Recapitulation

The purpose of this chapter was to find a neural substrate that can read the direction and speed of a trajectory - whether this is a sym-ax or a motion trajectory. We have approached this task using propagation maps and illustrated how speed can be estimated with them. What we have not addressed is how to read out direction. To perform that for sym-axes, one could use propagation maps with oriented connections. To perform direction selectivity for actual objects, the challenge of solving - or maybe circumventing - the aperture problem still remains.

A hardware implementation of this speed-estimating architecture required firstly the generation of spikes in response to motion, that could serve as input to the speed architecture (see bottom layer in figure 41). This already exists in form of a silicon retina (Boahen, 2002): it generates spikes in response to any object moving at almost any speed. Those spikes would feed into speed maps, which still had to be developed but could be easily derived from an implementation of the propagation map as presented in section 6.3.

The simulations in this chapter gave us the inspiration that shape could be detected analogous as speed is: a structure maybe stored as electric dynamics in a map, which would be activated only if the corresponding structure would 'run' across the map: the map would reverberate in some sense in response to the appropriate input structure. We continue this thought in later sections of the next chapter and we will also present a specific idea in chapter 10.

9 Neuromorphic Architectures: Pieces and Proposals

At the end of chapter 7, we have envisioned a hybrid categorization system consisting of two stages: in a first stage, a neuromorphic front-end encodes regions into sym-axes, and in a second stage, a computer vision back-end performs categorization with those sym-axes. But what if we intended to construct a pure neuromorphic system? It meant that we had to replace the computer vision back-end with a neuromporphic substrate that evolves and represents the categories. In what follows, we will firstly think about how one may continue such a neural evolement using the sym-axes. We then gradually turn towards alternative proposals.

9.1 Integration Perspectives

If one intended to continue with the sym-axes of the SAT, then a substrate had to be created that can integrate the sym-axes. Let us take the rectangle as an example (figure 44a). One may read out the individual trajectory pieces (numbers 1 to 5) by developing detectors that can sense the specific trajectory dynamics. The output of those detectors had to be added up to form the shape - in this case the rectangle. Alternative to this piecewise trajectory integration, one may create a detector that reads out the entire sym-ax of the rectangle. Either way, the rectangle represented had to be integrated with other regions of the entire object, as we did in chapter 5.

There are several approaches one could think of designing such *trajectory detectors*, as listed below. The first two ideas (1 and 2) have already been mentioned in chapter 8, when we aimed at reading out the trajectory of a L feature, whose sym-ax dynamics corresponds to a simple motion stimulus. The second two ideas (3 and 4) revolve about reusing and modifying the SAT.

1) Synaptic interaction: Trajectories may be decoded by synaptic interaction in the soma or the dendrite (section 8.1.2). Once, such single-neuron trajectory detectors had been designed, then a neuron would have to be developed that integrated the trajectory pieces to the representation (or detector) of the rectangle. Or, as mentioned above, there may be a detector for the entire sym-ax.

2) Maps: As mentioned in the previous chapter, a problem with synaptic interaction as just described in 1) could be the tedious alignment of orientations. To overcome this, one may use maps again: analogous to the idea of speed estimation, there would be maps reading out the trajectories wherever they would occur.

3) Stacking SATs: Instead of carrying out integration with single neurons or maps, one may also employ the SAT a second time. For instance, if one takes the sym-ax of a rectangle and squeezes it through

another SAT, then one obtains a sym-ax as shown in figure 44b. Each of those four, new sym-axes would represent three sides of the recangle (e.g. sym-ax 1 would represent sides a, b and c).

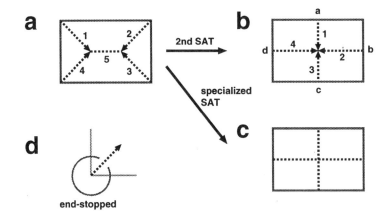

Figure 44: Integration of sym-axes. a. Sym-ax trajectories of the rectangle. b. Sym-axes of the rectangle's sym-axes (of a.). c. Sym-axes if only parallel structures would be detected. d. End-stopped cells to detect the space around sym-ax 'starting points'.

4) Specialized region encoding: There is a number of repeating structures in our environment, like parallel lines, certain curvatures or even entire shapes like the rectangle (see also (Lowe, 1987)). For such repeating structures, it may be worth to construct their own SATs. For example parallel lines would have their 'own' propagation map; the set of orientation columns would remain the same; and the symmetric-axis map would be accordingly wired to detect solely these specialized structures. The rectangle in figure 44a would so be represented by two sym-axis pieces forming a cross, of which each piece describes two parallel sides, see figure 44c.

As mentioned before, one type of region that has not been captured with the SAT is the encoding of outside space, in particular the silhouette of a feature, like the outside of a rectangle (section 7.1). In order to encode that outside region, one can analyze the spatial relationship of the sym-axes (section 7.3). One way is to employ *end-stopped* cells to signal Lone features. They are generally found in primary visual cortex and fire for the end-point of a line. This 'end-stopped' encoding can be applied for the beginning of a sym-ax as shown in figure 44d. Taking the rectangle as an example again, it would be described by the sym-ax as seen before, and in addition by four end-stopped cells, signaling the outside of each corner.

9.2   Position and Size Invariance

We now turn toward two aspects of the recognition process that we have only shortly mentioned in connection with Deutsch's shape recognition system (section 3.5).

When we take a first glance at a scene, the fovea may land somewhere, where there is not exactly the 'center' of an object: the object's position may be just slightly off the center. Still, recognition occurs without interupt and one refers to this recognition aspect as *position or translation invariance*. Likewise, the object may not be always of the exact same size but sometimes it appears a little bit bigger or smaller. Still, recognition is fluent and this aspect is termed *size or scale invariance* (figure 45a).

We have not mentioned these two recognition aspects before, because they do not appear when one employs list methods, as for example with the extraction of polygon features as done in chapter 5, or the analysis of sym-axes, if one used the hybrid category system only. One is apparently confronted with those two aspects if one attempts to construct a pure neural recognition system. For example, if one attempted to construct a neuromorphic system as outlined in the previous section, then those trajectory detectors had to be able to deal with these variances as well, because the evolved sym-axes would be at slightly different places in the SAM, depending on the objects exact location and size.

Both, position and size invariance, are generally believed to be solvable with a *pyramidal architecture* made of neuronal layers, whose receptive field sizes increase with higher levels (figure 45b; (Rolls and Deco, 2002)). This concept is suggested by the finding that receptive field sizes increase along the cortical hierarchy (chapter 3, section 3.1). The most prominent model pursuing this pyramid is Fukushima's Neocognitron (Fukushima, 1988). In such a pyramid, the position of an object is found by integrating its contours through higher layers. This may be unspecific however, because this integration represents a 'coarsening' in which structural information gets washed out.

An alternative to pyramidal contour integration is *contour propagation* (figure 45c): McCulloch proposed that a shape is propagated concentrically thereby scaling it in size (McCulloch, 1965). But he was not specific about a detailed architecture. We suggest one in figure 46a. It comprises a representation map (RM) and a propagation map (PM), both connected neuron by neuron in a spatially corresponding manner. The synaptic weight values between the two layers would correspond to the shape to be represented, e.g. the connections along the shape's contours would have a high value, the connections distant from the contours would have a low value. This weight matrix is illustrated by the gray shape in the representation map. Let us think through some scenarios how the RM would answer in response to certain input shapes: 1) If the appropriate shape is placed concen-

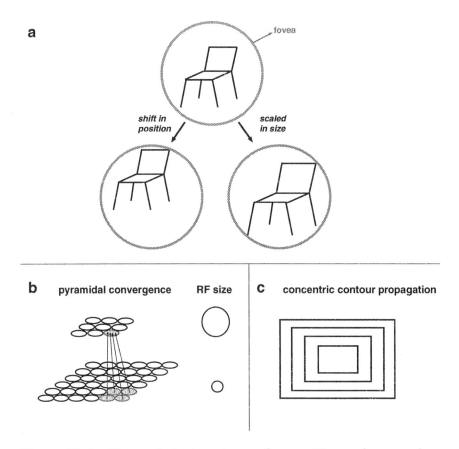

Figure 45: Position and size invariance of recognition and approaches to solve it. a. An object may be slightly off center (lower left) or slightly bigger or smaller (lower right). b. Pyramidal architecture with increasing receptive field sizes in higher levels. c. Concentric contour propagation as proposed by McCulloch (1965).

trically onto the PM, as outlined with i1 and i2 in figure 46b, then the outward- and inward-propagating contours would travel through the weight matrix simultaneously and the RM neurons would flame up simultaneously as well. 2) If the (appropriate) shape is placed with some offset or asymmetrically onto the PM, see corresponding rectangles i1 and i2 in figure 46c, then some sides will be activated before others and also partially only. Thus, the RM neurons would fire with some delay. 3) If a different shape, like a triangle or circle, was placed onto the PM, then the RM neurons would also fire at some point in time, but sequentially. In order to recognize the appropriate shape, one had to devise a recognition network that can cluster situations 1 and 2, but ignore situation 3.

The proposed architecture is certainly not useful in this exact format, but such thinking may be helpful for finding a more appropriate recognition architecture. This specific architecture would also neglect the idea of encoding space, something that we would not abandon after having made the valuable experiences in chapter 5.

A PM could also perform other transformations than just translation of contours. For example a map of ring-connected neurons arranged concentrically could perform a rotation of contours to some extent. That could allow for some *rotation invariance*, which would be helpful for recognizing objects that do not possess a certain orientation, like tools.

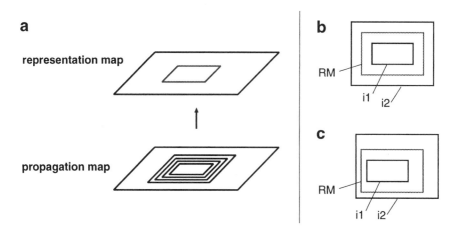

Figure 46: Propagation architectures for achieving size and position invariance. a. A template-matching architecture. b. Scenarios of concentric shape input: RM: shape represented in the RM. i1, i2: shape as placed onto the PM. c. Scenarios for offset and asymmetry.

It is difficult to foresee, whether the pyramidal or the propagation architecture (figure 45b versus c) solves the size and scale invariance aspect better. This had to be explored systematically. The optimal solution may possibly be a combination of both architectures. And it seems plausible to us that the visual system may bear such a combined architecture too: on the one side, the increase in receptive field sizes along the 'hierarchy' seemingly supports the pyramidal architecture (section 3.1); on the other side, most of the visually responsive cells are also motion selective, which ultimately may just reveal contour propagation dynamics.

Dynamic Representation In the above discussion, both architectures share the idea that the shape needs to be conveyed toward a representation expressed as a fixed set of neurons, which in turn have to be activated for recognition. Stated shortly, the input is made dynamic to move it into its correct slot, a fixed representation. But maybe it is this fixed representation that makes our search for position and scale invariance so awkward. How about making the representation dynamic? For instance, instead of representing the shape by a spatially fixed set of synaptic weights, it may be represented by a set of perpetually active (spiking) neurons in the RM, which would represent the shape in a structurally loose manner. After the shape is placed onto the PM and has triggered a number of waves, those waves in turn would stir up the activity in the RM and both the PM and the RM would transiently establish a strong connection between them, a 'locked' state in some sense: one may also term this a recognition by reverberation or attraction. Such dynamics have already been implemented in 'classical' neural networks (e.g. (Haykin, 1994) for a review). Our specific proposal here is that they would occur with waves that specifically encode space and that would solve some of the part-shape variability. A dynamic representation could also be of advantage for constructing the loose basic-level object representations we aim at throughout our book (section 2.5).

To what degree this size and position invariance is necessary is also something that has to be explored heuristically. To contrive the optimal architecture, it may be best to develop it with a system that performs saccades or attentional shifts that would bring the object into the 'exact' center of the focus.

9.3 Architecture for a Template Approach

In the section on 'Integration Perspectives' (9.1), we have listed a number of schemes on how to set the regions (sym-axes) into relation with each other to describe an object, which basically represents a structural description approach as pursued in chapter 5. However, in the end, any type of *integration* may be too *tedious* for the rapid association we look for (figure 2). Considering again that the visual system remembers pictures extraordinarily well (chapter 1), one may also ponder the possibility that some sort of template matching is at work in the real visual system. The idea exists since early on, but is often dismissed as too simple (Palmer, 1999). Nonetheless, it is this simplicity that could beat any sophisticated integration scheme. With the previously suggested architecture for solving size and position invariance, we have already moved toward some sort of template-matching scheme. We here give one outline on how to start thinking about a template architecture.

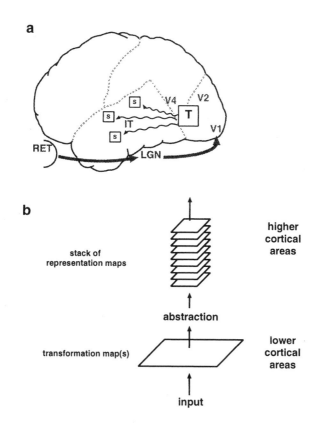

Figure 47: Recognition process with broadcast receiver principle. a. Lower cortical areas may perform a region-encoding transformation (T), whose product is sent across higher cortical areas to find the appropriate match (slot S). b. Neuromorphic architecture.

Let us assume that in the initial stages of recognition evolvement a number of transformations take place. These transformations would convert region space into an abstraction like a sym-ax, a set of waves, or a wavelet-like product, which would bear some structural variability independence (see 'T' in figure 47a). We imagine that such an abstraction could be stored as a whole. It maybe stored in a slot ('S') in a higher area of cortex, where it is appropriate for other associations, like other objects of a specific frame, or a set of motor actions that would be triggered if one intended to act upon the perceived object. Recognition matching would occur by propagation of the abstraction through the cortical medium until its corresponding slot is found. This idea can be classified as an instance of the broadcast receiver principle as it has been proposed by Blum. Blum's specific idea was that object properties would be integrated by coincidence of waves; our thinking, in turn, is that visual abstractions of whole objects or structures are propagated across the cortex.

In a neuromorphic system, this propagation process could be mimicked by projecting the output of the transformation maps into the chips containig the abstract representations (figure 47b).

9.4   Basic-Level Representations

In chapter 1 we have mentioned the existence of different category levels (section 1.5, figure 3), but throughout our chapters we have not been specific about the level we were talking about. We have an addendum on that.

'regular' chair   **office chair**   **bar stool**

Figure 48: More variability in the basic-level category chair. Chairs like the office chair or bar stool are only partially similar to the 'regular' chair that we have seen so far in this essay.

As mentioned in section 2.1 already, some basic-level categories possess a high structural variability, others a low variability. Categories with a low variability, like the category book, could be represented by a few regions or possibly by a single, abstract template. Basic-level categories with a high structural variability in contrast, likely require a larger, more diverse representation. Figure 48 should exemplify that point. The chair on the left, labeled as 'regular' chair, is the one discussed throughout this book, but chairs show also structures like the right two ones, labeled as office chair and as bar stool. They can be regarded as subordinate categories and are structurally only partially similar to the regular chair. The regular and the office chair share similar regions around the backrest and the seating area. The legs however are completely differently aligned, but share the property of having the 'legs' stick out into space. The bar stool is least similar to the other two. Only the bottom regions of the legs share some similarity with those of the regular chair. Given this variability, it is difficult to imagine that there is a single representational construct for this basic-level category. Rather, the variety of subordinate categories in this category has to be represented by several, distinct structural descriptions or maybe even different templates.

9.5 Recapitulation

In this chapter, we have reflected about how to construct a pure neuromorphic system - one that evolves and represents categories exclusively with networks. We have started by thinking about how to integrate sym-axes. We then have turned towards the recognition aspects of position and size invariance and discussed the pyramidal and contour propagation architecture that may solve these two aspects. We have expressed the intuition that representations may also be dynamic for example consisting of a set of waves. Furthermore, we have considered a template architecture that may store structures as a template, because a structural description scheme could be simply too sophisticated and expensive for rapid associations.

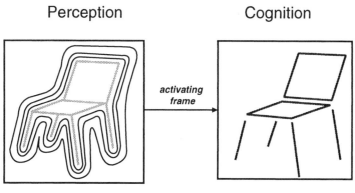

Figure 49: A perceptual representation maybe a collection of waves. Compare again to figure 2.

At this point, it is not clear to us yet, at what level exactly all these aspects and processes should take place in a recognition sytem and whether there is a specific architecture that can solve everything elegantly in a smooth manner. The following chapter can be understood as a step towards that direction. To conclude this chapter, we caricature this view with figure 49 (compare again to figure 2): it should express the thinking that waves can represent perceptual categories in some way.

10 Shape Recognition with Contour Propagation Fields

We continue the thought just discussed in the previous chapter, namely the idea that representation is some neural substrate that remembers propagating waves. In this chapter we present a network that can be regarded as an instance of this thinking. We demonstrate the operation of the network on simple shapes first and then discuss how it may be extended to basic-level categories.

10.1 The Idea of the Contour Propagation Field

In the search for a suitable network, we also aimed at a mechanism that would encode not only the inside but also the outside of a shape. The simplest principle that one could think of, is to remember the entire flow of inward- and outward propagating contours of a shape as is, meaning without the generation of a particular transformation. In order to represent the geometry somehow, one can look continuously at this contour flow through a set of orientation columns. By doing that, one observes a field of vectors as shown on the example of a rectangle in figure 50. Both, the inward- and the outward-propagating contours, contribute to this vector field, which looks akin to the optical flow field used in motion detection studies (Palmer, 1999). We call this vector field now the *contour propagation field (CPF)* (Rasche, 2005d).

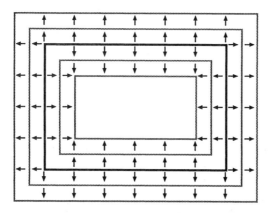

Figure 50: Describing shape with the contour-propagation field (CPF). Looking at the propagation pattern through a set of orientation columns generates a vector field analogous to the optical flow field used in motion detection studies. We call this flow field now the contour-propagation field.

Next, we need a neural substrate that is able to remember this CPF.

For example, one can make use of those motion-sensitive maps discussed in chapter 8. If the connectivity of such a motion-sensitive map is altered in such a way that it corresponds to the CPF pattern, then the motion-sensitive map would react only to input of corresponding contour flow (or CPF). We now call such a motion-sensitive map, with a specific connection pattern bearing a CPF, a shape map (SM). For each shape, a separate shape map is created. Recognition would occur by feeding the contour flow across each shape map. Only, the map with corresponding CPF will generate a vigorous response, and that map response serves as recognition signal. This comparison process is in some sense a template matching system. We therefore call it CPF matching, or CPFM.

10.2 Architecture

An architecture for a CPFM can be constructed with the components discussed in chapters 7 and 8. Figure 51 shows the principal architecture for both solutions. The input shape (*input*) is fed into a propagation map (PM) and analysed by a set of orientation columns (OCs). On top of that, a set of direction-selective columns (DCs) determine the direction into which a wave propagates. (The use of such a direction-selective cells would of course be also beneficial for the SAT, but has not been modeled yet). The direction selectivity can be obtained by cross-correlating the input of two neighboring orientation-selective cells, very much according to the principles discussed in the introduction of the motion detection chapter (number 8). At the very top of the architecture is the set of shape maps (SM), each representing an individual shape. Each shape map receives input from the direction-selective columns.

Shape map A shape-sensitive map is basically equivalent to a propagation map, except of two modifications: 1) The connection between two neurons are modeled as two unidirectional connections, see figure 52. 2) Most of its connections are turned off, and only those are turned on, whose direction corresponds to a local direction of the CPF. For example, the connections around the center neuron shown in figure 52 are turned off (grey) except those two pointing toward the right (black): this local piece of shape map, will prefer contour propagation toward the right only. To learn a novel shape, the shape is squeezed through the architecture and the evolving CPF is 'burned' into an 'unwritten' shape map, which then becomes part of the system. Once the novel shape map has been burned, it basically consists of lines of connected neurons.

The architecture suffers from the problem of a coarse connectivity, in particular the SM. Specifically, a direction column provides 16 distinguishable directions, whereas a SM neuron has only 8 neighbors

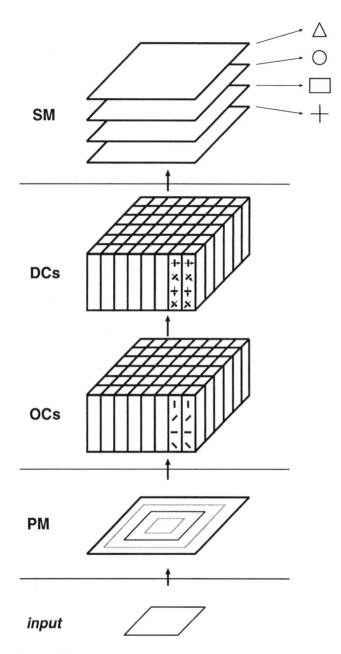

Figure 51: Architecture for a CPFM. From bottom to top: the input shape (input) is dipped into the propagation map (PM), where it propagates inward and outward (gray outlines). The orientation columns (OCs) and the direction columns (DCs) determine the contour-propagation field (CPF) which is then fed into each shape map (SM).

and thus 8 directions, which is too coarse for representing simple shapes distinctively. For the present study, the problem is solved by using two propagation layers for each shape map, each one propagating the input from 8 different directions.

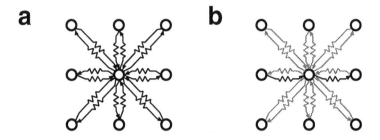

Figure 52: a. Reciprocal, unidirectional connections in a propagation map for one unit. b. Only one direction of the 'directed connections' is turned on (black), the others are turned off (gray): this map will prefer contour input moving towards the right only.

10.3 Testing

We have simulated this architecture with integrate-and-fire neurons, except of the block of direction selective cells, which has been computed by traditional bit-wise (computer) correlation for reason of simplicity. To determine the recognition signal, the entire spiking activity of each shape map is observed by summing it up and plotting it against time. This sum is now called the population activity. Five shapes were used for testing, see figure 53: a rectangle, a circle, a triangle, a cross and a shape consisting of the superposition of the rectangle and the cross, which we now call square board. The 1st column is the centered shape, which was used for learning - for burning the CPF into the shape map. The 2nd column is a translated version of the shape, shifted to the lower right by 5 pixels in each dimension, which represents about 10 to 12 percent of the shape's width and height. The 3rd column is scaled version of the (centered) shape, made smaller by 6 pixels in each dimension. The 4th column is the (centered) shape with a 'disturbance', a straight line extending from the right side of the image center to the upper right. The 5th column contains the shape made of a dotted contour, an input that is hereafter called fragmented. Figure 54 shows the CPF of two shapes.

The ability of a shape map to identify was tested with its own centered, shifted, scaled (smaller), 'disturbed' and fragmented shape. The ability to discriminate was tested by placing the other, centered shapes on it.

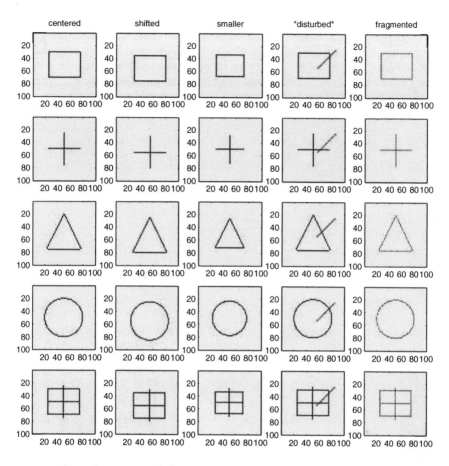

Figure 53: Shapes used for testing. From top to bottom: rectangle, cross, triangle, circle, square board. The 1st column shows the centered shapes. The 2nd column shows the translated shapes, which were shifted by 5 pixels into both axis directions (shifted to the lower right), which is about 10 to 12 percent of the shape width. The 3rd column is a down-scaled (smaller) version of the (centered) shape. The 4th column shows the (centered) shape with a 'disturbance', a straight line extending from the right of the image center to the upper right. The 5th column is a dotted version of the shape, called fragmented.

Figure 55 illustrates the population activity of the shape maps in response to various stimuli. The identification responses are discussed first.

The response to its 'centered' shape, called the signature now, is denoted as a thick, solid line: it rises steeply and stays well above the

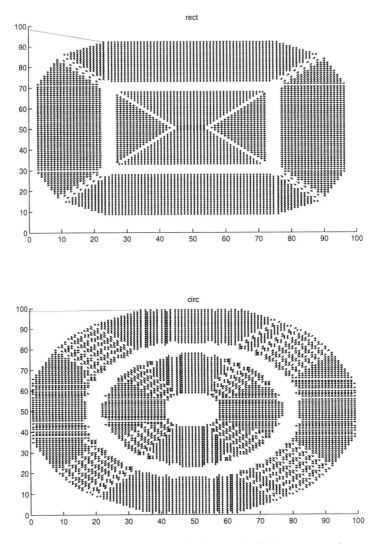

Figure 54: Contour propagation field (CPF) of the centered rectangle (top) and circle (bottom). An inward-pointing and outward-pointing vector field is present.

response activity for any other identification response. Its amplitude is determined by the length of the shape contour. It may increase or decrease during the propagation process depending on whether contours cancel each other out (e.g. square board) or whether they are only growing (e.g. cross). Part of the fluctuations are due to the aliasing problem (coarse nature of our network).

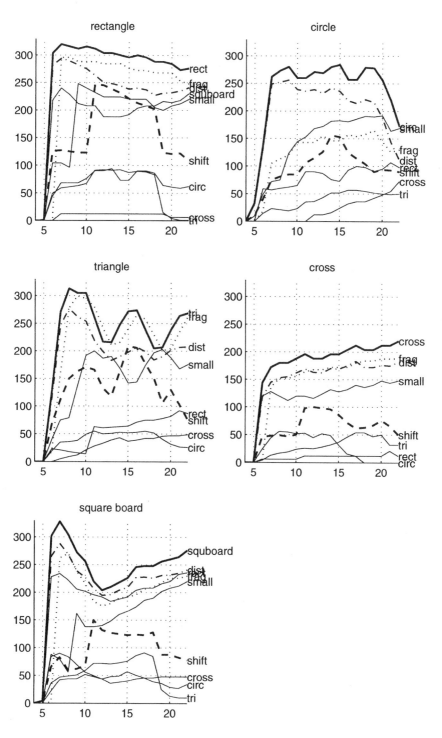

Figure 55: Population activity during recognition evolvement for the five different shapes. X-axis: time. Y-axis: spike activity of the entire map (population activity).

The response of a shape map to its 'disturbed' shape is shown as a thin, dash-dotted line, which also rises steeply but then runs below the signature: The 'disturbing' line causes a subtraction because it 'steals' a piece of area.

The response to the fragmented shape starts with a delay (of one time step) because it takes a short while for the PM to fuse all the contour pieces to one continuous (inward- or outward-propagating) contour.

The response to the down-scaled (smaller) shape starts immediately but reaches its peak stepwise because it takes a short while until both inward- and outward-pointing vector fields are covered, except for the cross shape.

The response to the shifted shape - plotted as thick, dashed line - is slower and reduced compared to the other identification responses. It also reaches its peak stepwise for the same reason as for the smaller shape. A translation by 10 pixels was also tested but those responses were not significantly different anymore from the discrimination responses.

The discrimination responses - the response to other (centered) shapes - are plotted as thin, solid lines. All those responses are mostly below and sometimes minimal compared to the identification responses. If, however, two patterns are similar in their exterior space, as it is the case for the rectangle and the square board, then they may have overlapping identification and discrimination response.

10.4 Discussion

The use of two propagation maps for each shape (to circumvent the aliasing problem) is somewhat awkward. A better solution may lie in the usage of synapses representing the range of orientations for determining direction, instead of an actual propagation map. Nevertheless, the simulation results so far suggest that any shape can be easily distinguished from any other shape, as long as they are not structurally too similar. The architecture is therefore good for simple 'shape categorization': a learned shape can appear fragmented, scaled, disturbed, shifted to some extent and still be identified; likewise the learned shape may have minor deformations or slightly different structure, it would still be recognized and hence categorized. Consequently, the architecture is less suitable for an exact identification and it could less efficiently discriminate between subtle shape differences, like the ones shown in figure 56. We discuss improvements in a subsequent paragraph.

The CPFM system possesses all the desired features we dream of a potent neuromorphic visual system:

1) The system learns the shape by itself. This is a feature we have not sufficiently pointed out yet and will be further discussed in the

subsequent subsection.

2) The system shows some size and position invariance to its input shape.

3) The system is enormously robust to 'noise' sources like contour fragmentation or additional contours.

Features 2 and 3 are the result of two properties of the system. One is the propagation of the input in the propagation map (PM), and the other is the wide-spread encoding of the shape, the CPF. Thus, this network is an answer to the problem formulated so well in Palmer's book ((Palmer, 1999), p.86-88): How is that a shape can be recognized although its contour image is always incomplete and varies with different illumination? The answer may be in a network as described here: A network that encodes the region and uses wave propagation to do so.

Comparison to the SAT   How does the CPFM compare to the SAT? Both mechanisms are region-encoding mechanisms, but they differ significantly in their specific evolvement and representation.

- Recognition duration: The time it takes to signal the proper shape is relatively immediate for the CPFM system because the CPF representation is wide-spread in some sense and thus allows for a quick response. In contrast, the SAT requires more time, because the entire region - or at least a substantial fraction of it - has to be explored by the traveling wave in order to form a least part of the sym-ax.

- 'Noise' sources: Contour gaps: Both mechanisms deal well with contour gaps. In case of the SAT, the gaps are closed by the propagation map (PM) only. In case of the CPFM, it is also the propagation map, but also the wide-spread CPF representation that circumvents the need for closing all gaps. Additional contours: The CPFM is very robust to noise sources like the additional line. The SAT has a larger problem with that, because additional contours can substantially alter the outcome of the SAT process and generate sym-axes that are only remotely similar.

- Size and position invariance: The SAT does not really address this aspect because it does not contain a substrate that stores the sym-axes. Using a computer vision back-end however, a list-approach, would show a large degree of size and position invariance as it is the case for any list-approach, like the one carried out in chapter 5. In contrast, the CPFM shows a limited size and position invariance due to the somewhat fixed representation.

- Representation: The SAT only encodes regions that are 'engulfed' by contours, whereas the CPFM encodes any region - the inside as well as the outside. The output of the SAT is graspable for a list approach, whereas the present CPFM is only suitable for a pure neuromorphic approach. Both processes likely have difficulties to efficiently distinguish between subtly different shapes like the ones in figure 56.

- Structural variability: both processes are robust to structural variability, although we have not specifically shown that for the CPFM.

CPFM improvements   There are several directions one may pursue to improve the present CPFM system. Firstly, it may be useful to distinguish between straight lines and arcs. To extract straight lines, one would employ larger receptive fields - as it is possibly done in the real visual system by a local-to-global recognition evolvement (section 3.1). That would lead to the employment of a set of different maps, each for a different size of receptive field. Curves would then be represented in maps with small receptive-field size, and straight segments in maps with large receptive-field size. Another direction of improvement could be to represent shape on different scales (coarse-to-fine) as Koenderink and vanDoorn proposed (Koenderink and van Doorn, 1986). This may indeed work with some objects and even textures, but for the shapes shown in figure 56, this could be of limited use because at a coarse scale, those shapes appear even more similar.

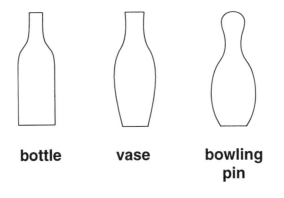

**bottle   vase   bowling
                                    pin**

Figure 56: Similar shapes. The CPFM, but also the SAT, may have difficulties to efficiently distinguish between these shapes. That may be solved by employing different receptive field sizes (local/global) or different spatial scales (coarse/fine) or by inventing a differentiation mechanism.

Extension to Basic-Level Objects   The CPFM is able to represent simple shapes only. Basic-level categories, as we have used them in chapter 5, can probably not be efficiently described by a single CPF. There is likely too little position and size invariance and too little viewpoint independence, all of which relates to each other somewhat for this type of representation. Thus, the CPF idea had to be elaborated in some way. For example, if the tolerance for size and position invariance would be larger, then it may suffice to represent a basic-level category by a set of CPFs, one for each region as determined in chapter 5 for instance. Stated differently, each of the features that we have determined in chapter 5 would be expressed by some sort of CPF.

Broader Implications   If a more potent CPFM system can be constructed, one that can perform basic-level categorization and that would recognize its input as easily and distinct as the present CPFM does, then that would have a number of broader implications.

1) The system can potentially recognize serveral objects at the same time. Imagine that the above simple CPFM system is stimulated with a square containing a circle inside. The shape maps for both shapes would lit up, without little interference during the recognition process. Thus, objects that overlap each other, could easily activate their representations simultaneously.

2) The system has the potential to operate in a complete bottom-up manner, meaning it does not need any 'top-down' control, frame interaction or other representational control to understand noisy scenes or scene parts, unless the object or scene is vastly degraded in some form. Thorpe has argued that the visual system has to operate purely bottom-up because of its tremendous processing speed (section 3.3). CPF matching could explain that speed.

Biophysical Plausibility   Because we regard the CPFM as an instantiation of the thoughts in the previous chapter, we do not particulary explain the plausibility of this system, but comment on the speed of recognition only. In the CPFM system, the waves travel only a relatively short time until the proper shape has been recognized. Because of this short recognition time, the waves do not have to be as fast as for example for the SAT, in which a wave has to travel some portion of the region to establish the sym-ax. Hence, the fast part of recognition may be only the propagation to different shape maps in cortex, but the wave that actually would stimulate a shape map did not have to be particularly fast.

10.5   Learning

The shape-recognition system has a formidable property, which is of great importance for the construction of a self-organizing neuromorphic system: it learns. We have not talked about this property before, because we have focused on finding an effective description. Once, such an effective descriptor is found, then it makes sense to develop a learning system. The invent of the CPF idea brings us closer to this goal, because the CPFM system represents effectively and is able to learn simple shapes in a single pass. It may be noteworthy that learning in the CPFM system requires an 'unwritten' map. This is in contrast, to traditional neural networks which do not increase their number of neuronal units but retune their existing weights to incorporate the novel shape.

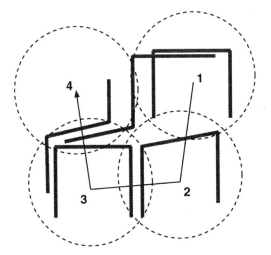

Figure 57: Learning the representation of a chair. Ideally, one would create a neurmorphic system that learns the significant regions by itself by hopping from one region to another and thereby gradually acquiring the template-like region representations that form the loose category representation.

If one would be able to construct an expanded CPFM system, that can deal with basic-level categories - as envisioned above for instance -, then that would mean that there needs to be a process that detects these regions in a novel object. This process may consist of a saccadic-like visual search to acquire the various regions. For example, the system would jump from region to region and generate a template for each region (figure 57). To find the center of a novel region, one could preform a simplified or modified SAT on the object, although not all regions can be found with the SAT due to its drawbacks.

Differentiation If an efficient learning system can be found, then a subsequent challenge would be to construct a system that learns to differentiate. For example, if some input may appear very similar to a learned (stored) shape, but has to be distinguished from that one, then the system has to generate a novel shape map or modify that stored one. This would occur for instance, if the system tries to form subordinate categories, like kitchen chair or recliner seat (section 1.5). One may exercise this differentiation step on the example of simple shapes first, and then extend it to basic-level categories.

10.6 Recapitulation

We have described simple shapes by remembering their contour prop-agation pattern as viewed through a set of orientation columns. The resulting vector field - called the contour propagation field, CPF - de-scribes the inside and outside of shape (expressed in figure 58). A neu-ral substrate storing this CPF can properly recognize learned shapes, even if they are scaled in size, translated in position and degraded in various manners. The system learns a novel shape with a single 'shot' using an 'unwritten' map for each new shape. The CPF matching sys-tem can deal with simple shapes, that are structurally not too similar. We have mentioned several possibilities to extend the system, like op-erating on different axes (local/global) or by creating a differentiation mechanism. If such a system can be developed for basic-level objects, it has the potential to work purely bottom-up despite the persistent presence of noise in the contour image, and it has the potential to rec-ognize several objects simultaneously, even if they partially overlap.

Perception   Cognition

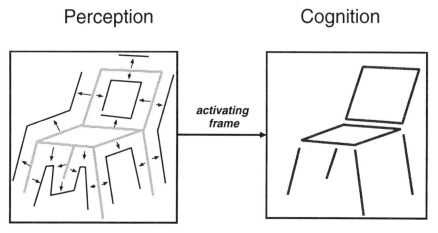

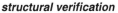

fast categorization   **structural verification**

Figure 58: The idea of using a CPF for representation is expressed with a set of arrows indicating the direction of waves.

In search of an expansion to a basic-level categorization system, one would firstly attempt to succeed on line-drawings like the ones in chapter 5. Such line-drawing objects could include a substantial amount of variability - more then the ones in chapter 5 -, because the present CPF system is already able to deal with a significant level of variability. Ideally, in a second step, one would design a learning system that can acquire the necessary regions by a saccadic visual search. Then, one would elaborate the system to perform on gray-

11 Scene Recognition

Scenes are made of basic-level objects. To determine a scene's content, it seems obvious to firstly recognize its individual objects sequentially and then to construct the 'understanding' of the entire scene. Our belief is that once the categorization process is properly built, then it can be readily applied to scene recognition. The goal in this chapter is to argue for that belief but also to suggest that scene understanding is more instantaneous than the commonly believed sequential build-up from object information.

11.1 Objects in Scenes, Scene Regularity

If an object is placed into a scene, then its silhouette (or outside) space is often dissected by structure that is part of the scene's global structure. For example floor contours intersect with the lower section of a piece of furniture. Or furniture that is close to each other masks out each other's silhouette features. But despite these 'background contours' there is plenty of typical silhouette space around an object that could allow a network using region encoding - as we have pursued it in the previous chapter - to perform proper recognition. What would happen if the scene was densely cluttered with objects? It certainly took longer to understand the scene and to comprehend each object. Two examples come to our mind where this is obvious: 1) In a garbage dump, objects are intermingled - and also seen from unusual viewpoints. 2) In certain bars (theme bars), it is popular to decorate the walls and ceilings with diverse objects. In both cases, it takes a long-lasting visual exploration to recognize each object. Biederman has elegantly shown that point in the context of scene understanding (Biederman, 1972; Biederman et al., 1973): If objects are placed in unusual locations, or if scenes are jumbled, it takes the human observer some while to understand the contents of a scene or to find a typical object. Thus, objects possess a *typical scene context*. One can take the conclusion a step further and claim that scenes as a whole possess a 'regular' structure like basic-level objects do and that they show the same types of variabilities as we described for objects, see section 2.1. And because scenes are so regular, most objects will have at least some of their silhouette region always 'available' for rapid analysis.

11.2 Representation, Evolvement, Gist

How are scenes represented? Because scenes consists of basic-level objects, it is commonly believed that scenes are best represented by them. This is the same thinking as for the description of objects by parts. And because a scene has apparently more structure than an object, it is believed that scene Evolvement requires a set of mechanisms to decompose the scene into its constituent objects. Such segregation mechanisms have been termed for example perceptual grouping, region segmentation or figure-ground separation (Palmer, 1999). Such an elaborate reconstruction process may be in contrast with the speed, with which humans can understand the gist of scenes. Gist perception can occur within a glance (section 1.2), and it creates a rich *sensation of impressions*. There is a large debate on how much is perceived within a single glance and how much of it reaches our awareness or consciousness (Koch, 2004). And there have been efforts to find a specific evolvement type for gist perception, some of which we have already discussed. For example, we have already discussed the local/global debate (page 25). Likewise, followers of the channel theory have wondered, whether an image is firstly processed from a fine (high frequency) to a coarse (low frequency) scale or the opposite way. Specifically, Schyns and Oliva proposed that for short image presentations a coarse to fine scale analysis takes place. Yet, after further studies this conclusion seemed not to be that firm anymore: gist perception can also occur from fine-to-coarse depending on the task (Oliva and Schyns, 1997). Another way to formalize gist perception is to separate scene information into foreground and background: Some psychologists claim, that we understand a scene by perceiving a few key-objects (foreground) in a scene and then to conclude to the embedding content (background). However, the majority believes that scene analysis starts with the background information and then determines what objects there are in the scene (debate reviewed in (Henderson and Hollingworth, 1999)).

The multitude of suggested mechanisms and the diversity of experimental results may also be simply the results of a flexible recognition evolvement that allows to analyze the scene the easiest possible way. The scene may actually drive the mechanisms (Rasche and Koch, 2002). But our preferred viewpoint of scene recognition is to regard a scene as a basic-level object in some sense. For example, room scenes share a certain structure as opposed to for example landscape scenes. And because they possess a lot of regularity as argued in the previous section, they possess the same types of structural variabilities as objects do (section 2.1). We therefore propose to apply our idea of loose representations to scenes as well: as for basic-level objects, there may be some sort of pattern, that can be expressed with a set of templates. Or a scene maybe can be even *classified* based on its regions. The idea of mere classification has already been put forward by Duda and Hart (Duda and Hart, 1973).

We attempt to clarify our idea of equating scenes with basic-level objects on the example scene in figure 59a. Assume that one enters a room scene as shown in figure 59a. With the *first glance* into the room scene, one would categorize the object in the fovea, in this example the painting on the wall (figure 59b). The region (or 2D space) around the painting is a characteristic region found with most paintings: they tend to be isolated on the wall - for reason of aesthetics. The surrounding, peripheral structure (shown in gray) has been perceived as well with this first glance, but it has not been specifically categorized: it merely represents typical structure that triggers - together with the categorized object - the corresponding frame, in this case the *frame 'room scene'*. Both, the categorized object and the peripheral structure, can be structurally variable. With this activated frame, much of the peripheral structure can already be guessed, even if the contours have not been completely extracted. For example the rectangle to the left is likely to be a window, the structure on the right is likely to be a standing lamp, the structure below is a desktop or table. If we had only this first glance of a room scene, like in Thorpe's experiment (section 3.3), it would already suffice to determine a number of objects in very short time.

Thus, scene categorization, or gist perception, would occur based on the categorized object in focus *and* the structure in the periphery. No detailed or sequential reconstruction would have occurred: it is merely the typical scene pattern that has triggered a host of associations, a frame. In some sense, our idea is not far removed from the Gestaltist's idea that a scene is swallowed as a whole.

11.3 Scene Exploration

Following gist perception, the visual system begins to explore the scene by choosing different spots in the image for further analysis - it starts performing saccades to these spots. The selection of those subsequent spots is possibly guided by the activated frame and this guidance may be called hypothesis-driven or top-down control. But there may also be bottom-up or also called stimulus-driven aspects of the visual scene that guide the saccadic selection process. For instance, Koch and Ullman have proposed that a *saliency map* exists somewhere in a high visual area (Koch and Ullman, 1985) that collects the most conspicuous spots of the visual image in a purely bottom-up manner, which in turn are selected based on the degree of saliency. Itti has modeled this idea using an interaction of orientations, colors and blobs taken at different resolutional scales (Itti and Koch, 2000). This pure stimulus-driven exploration may occur when there is no particular motivation during scene exploration. On the other hand, if a human searches for a specific object, then a visual search occurs, which may be hypothesis-driven.

There may also be intermediate forms of those two extreme types of scene exploration. What exactly happens during exploration is difficult to elucidate and is further complicated by the fact that there are also attentional shifts between saccades. Let us assume that the

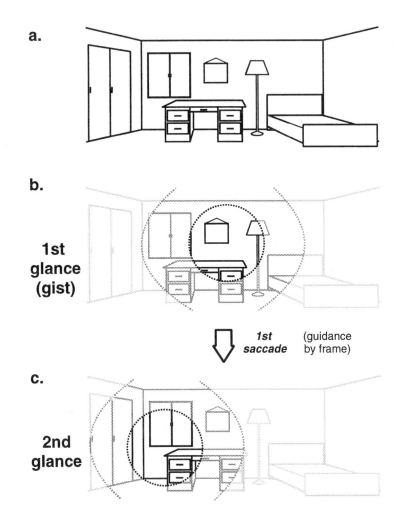

Figure 59: Scene recognition. a. The room scene. b. The 1st glance may land on the painting triggering the frame room scene and giving the sensation of gist. c. The 2nd glance is part of a visual exploration already.

selection of the next spot has taken place somehow and that the first saccade brings the focus onto that specific spot (figure 59c). This *second glance* may investigate the windows sills and may be already part of a scene-exploratory process. Another part of the desktop and the window is now in focus, as in the first glance already, but neither object has been necessarily fully categorized, nor is their structure verified: it could be an illusion! Thus, a scene would be explored using only those loose and abstract perceptual category representations we were hunting throughout the book (figure 2, left side; figure 61).

Even if one had seemingly much understood of the scene with these two glances, there are still big junks of it, that are not actually recognized. We think that this is the reason why phenomena like change blindness occur (Rensink, 2000; O'Regan et al., 1999): there is just way to much detail in a scene that could be comprehended, even after a long visual search (chapter 1). This point has already been made by others in one or the other way, for example (O'Regan, 1992; Koch, 2004). But what we specifically would like to emphasize is that the scene is understood with perceptual category representations that are loose and abstract, and that this is the major cause for those observed phenomena.

In summary, scene exploration is not so much about the systematic, stepwise recovery of its content, but is a process consisting primarily of making rapid associations based on a small fraction of the structure perceived in a scene. The first rapid association triggers a frame, the gist perception, which in turn is the guidance for the subsequent eye fixation causing another rapid association, and so on.

11.4   Engineering

How would one then approach the construction of a scene recognition system? As we have tried to argue in the previous sections, recognizing a scene is like recognizing a basic-level object. Ergo, one has to seek the neural substrate that is able to deal with structural variability. In other words, it boils down to the construction of the basic-level categorization process as we have pursued in the chapters presented in this book. Once such a basic-level categorization process has been constructed for the line-drawings in chapter 5 for example, then one may test it on scenes like in figure 60. Although the ultimate goal is certainly to have a system recognizing objects in gray-scale images, tackling the structural variability in real-world scenes at once may be just too overwhelming. It may therefore be sensible to develop a system that is able to deal with a reduced variability as we did in chapter 5. The room scenes in figure 60 are literally identical with regard to scene content but contain structural variability that can still not be dealt with in an elegant manner by any scene recognition system. Many representational issues can be explored with such simple scenes, without dealing with the full-blown variability existent in real-world scenes. We think that the translation of a line-drawing analyzing system to a real-world (gray-scale) analyzing system could then be systematically carried out.

One may not intend to wait until the categorization machinery has been thoroughly constructed and one may already plan to develop a search system that tries to find and identify simple shapes in scenes.

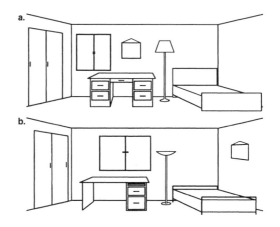

Figure 60: Two similar room scenes (a and b). There is still no elegant network that can deal with that type of variability.

Since we have a shape-recognition system that is capable of identifying simple shapes (the CPFM system, chapter 10), one may test this system on real scenes and gain possibly valuable experiences. For instance, one could design a saliency process aiming at regions using a crude form of the SAT that locates the center of shapes. The system would determine simple structures by saccading to those center points and by sequentially applying the CPFM system to identify the shapes.

11.5 Recapitulation

Scenes have a very regular structure and the representations one has about these scenes incorporate this regularity in some way. It therefore makes sense to treat scenes similar to basic-level objects: there is a common structure amongst scenes of the same category and that common structure - including its characteristic region of course - is stored in some way; and when it is perceived, it triggers the sensation of gist perception. Exploring a scene by saccades is making associations between perceived fragments of the scene; scene exploration does not require detailed reconstruction. We have suggested that one could begin building a crude scene-exploration system using a process to find salient regions, like a modified SAT, and employing the CPFM system to determine the shape of the regions.

12 Summary

Our approach to the construction of a visual system focuses on an implementation of the categorization process. The basic assumption is that if an architecture can be developed that performs this process, it will lead to an extension into scene recognition and to a refinement required for recognition in low-resolution images (section 2.5).

12.1 The Quest for Efficient Representation and Evolvement

The goal was and still is to find category representations for canonical views, that are distinct and easily evolvable. These category representations have to be loose in order to be able to deal with the structural variability existent across category instances (chapters 1 and 2). Structural variability can be classified into part-shape variability, part-alignment variability and part redundancy (chapter 2, section 2.1). Because it is difficult to contrive such a loose representation beforehand - or the architecture that evolves and bears such representations -, we have approached our goal in an exploratory manner. Figure 61 summarizes this quest, whereby the top two chair drawings (a+b) represent two popular approaches to object description and the bottom four chairs (in c) summarize the progress we have made.

The chair schematic in figure 61a caricatures the part-based approach in which an object is organized into its constituent parts. We have not followed this type of representation, because it focuses on object parts solely, which may not be able to deal with the structural variability and because a part-based interpretation rather represents an end-stage of cognition (chapter 2). The chair in figure 61b illustrates the neural network approach ('NN') in which in a first step the object is dissected into its local orientations, and in a second step these orientations are gradually integrated. We have not followed this line of reconstruction either, because it is structurally to unspecific (chapter 3). Our approach aims roughly at an intermediate level between these two descriptions: Similar to some of the neural network approaches, we try to find a fast match between image and representation, but it should be structurally more specific, yet not as elaborate as a part-based approach. The top left chair in figure 61c (labeled 'space') summarizes our computer vision studies of chapter 5. In these studies we have identified region (or 2D space) as a crucial component of representation: region encoding helps to bind contours and to form distinct representations. The regions that were used as features can be described as surface and silhouette regions. We then translated the idea of encoding space into a neuromorphic language, thereby using traveling waves as a means to sense space. Specifically, we employed Blum's symmetric-axis transform, that encodes regions

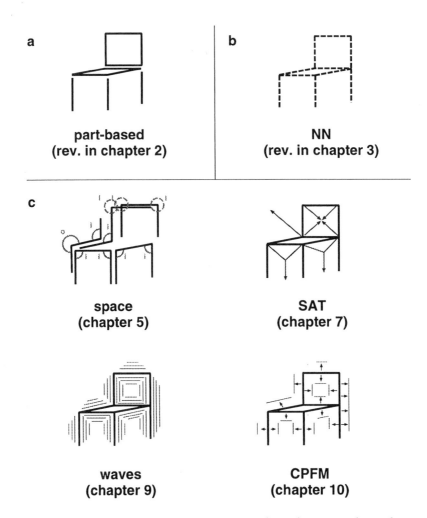

Figure 61: Chair representations. Part-based reviewed in chapter 2. Neural network ('NN') reviewed in chapter 3. Representation with space (chapter 5). Encoding space with the SAT (chapter 7). Representation by waves (chapter 9). Representation and evolvement with the CPFM system (chapter 10).

into sym-axes (top right chair in figure 61c, labeled 'SAT'; chapter 7). The traveling waves of our propagation maps have the characteristic of sealing gaps of the extracted, fragmented contour image, and they also smoothen out some of the structural variability. Finally, the bottom left chair in figure 61c caricatures the ideas we have proposed in chapter 9: we envisage that structure maybe represented as a set of waves somehow, thereby possibly solving also position and size invariance to some degree. A structure may trigger specific and distinct waves which are remembered by a neural substrate sensing the emitted waves. The bottom right chair expresses the idea of contour-propagation-field matching followed in chapter 10, which can be regarded as an instantiation of the ideas formulated in chapter 9. If this mechanism can be extended to represent objects then it had the potential to run completely 'bottom-up' and to run blazingly fast.

A melange of various mechanisms?   After having tried out several shape and object descriptions, invented a new one, discussed some more and considered even mixtures between them, it dawns me that there may not be a single method or mechanism for representing visual information. All the proposed mechanisms have their benefits and drawbacks, see Palmer (1999) for an elaborate comparison of some of them. However, we regard the encoding of space as a crucial fundament of visual description. Mounted on top of that space-encoding mechanism, could reign a structural description approach, a classifier approach or maybe another step of space encoding. In our studies we have sensed space with traveling waves. One alternative possibility would be to carry out some sort of channel approach, although an exact transform of that type is not desirable, because the encoding should allow for categorization. That means it has to be able to deal with structural variability. Another alternative maybe a system as Deutsch proposed it (see section 3.5;(Deutsch, 1962)). But as we just noted above, one may have also to consider that basic-level categorization is simply not solvable with any single type of space encoding and structure representation, but can only be performed with a mixture of several types. Evolution may have hacked together those different types of mechanisms and representations.

Object, scene and texture all the same?   We have primarily focused on object recognition and have marginally touched the issue of scene recognition. We proposed that objects and scenes may not be so different in their representation after all. One may take this a step further and suggest that textures may also be described like objects and scenes: by some sort of loose structure. Followers of the channel theory have already proposed that there may be a single type of encoding, but that approach has not generated any specific example of shape or object description - and as argued above, we anticipate that the solution to the problem maybe somewhat more elaborate than just a single transformation. Although there may be several mechanisms at work to represent structure, they may all serve to describe any type of visual information, whether it is texture, object or scene.

Perception-Action   An animal reacts to its visual percepts, and sometimes it does it as fast as it can (Bruce et al., 2003). Therefore, per-

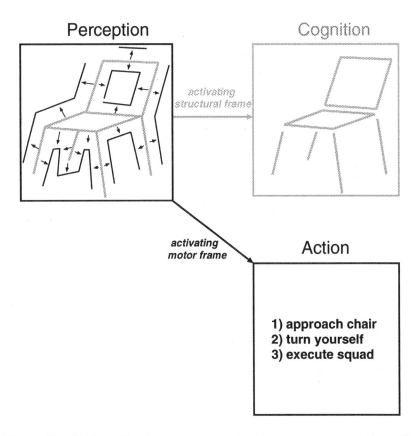

Figure 62: Linking the loose perceptual category representations directly to action. There is no particular need to reconstruct everything during interaction with our environment. The loose representations could be directly associated with 'motor frames'.

ception is likely directly connected to action, meaning there is a direct association. We think that not much has changed about that principle throughout evolution: even if the percepts are basic-level category objects as in humans, and even if humans have the possibility to respond in many more different ways to an object than simple animals. In our daily lives we rapidly and fluently carry out innumerous perceptual decisions that lead to instantaneous actions. This swiftness could require a direct association between percepts and action. We therefore envision that the perceptual category representations we aim at, do directly trigger motor actions. For instance, if we see a chair and intend to sit on it, this would trigger a motor frame loaded with actions like 'approach', 'turn yourself', 'bend knees', and so on (figure 62). Or in response to seeing an apple, a motor frame is triggered with actions like 'grab with fingers','move to mouth','bite'. Neither the chair, nor the apple has to be structurally verified during visual recognition. In short, perceptual category representations would directly trigger motor frames. A structural verification would only seldomly occur because there may not be any particular need for it.

12.2 Contour Extraction and Grouping

To extract contours we have used a single neuronal layer with local connections and adjustable spiking thresholds or spike latencies. The resulting contour image looks like a contour image obtained with a computer vision algorithm. The contour image is sufficient for encoding the regions which are necessary to perform a perceptual categorization. One may seek to improve the contour image by applying perceptual grouping processes that are either purely computationally motivated, as Marr did (Marr, 1982), or that are motivated by psychophysical studies on contour grouping. There is a good amount of psychophysical work that may be easily applicable for that purpose (e.g. (Hess and Field, 1999; Bruce et al., 2003)). Such aspirations should not however result in the pursuit of an immaculate contour image. As we have already pointed out, there is enough region information in the contour image as it is obtained with our retina circuits. The long-term effort should be rather spent in finding efficient region and category representations.

12.3 Neuroscientific Inspiration

Much of current computational neuroscience has focused on unraveling the spike code that the brain maybe using (Dayan and Abbott, 2001; Gerstner and Kistler, 2002; Koch, 1999; Trappenberg, 2002). Given the blazing speed with which the brain operates and given that neurons fire at a low frequency (sections 3.2-3.4), it may well be that the brain does not use a spike code at all. Because the computations in this book are done with waves, the viewpoint that one therefore may adopt is that spikes appear as part of a wave: a wave propagating through the neuron, or a wave triggered by the neuron. With waves we have solved tasks like contour detection, contour binding and speed estimation, whereby the specifics of the implemented waves are variable: To signal contours, the wave is a charge-propagation mechanism; to encode space - or to bind contours -, the wave is an actively propagating wave; to estimate speed, the wave is inert and responds to preferred speeds only. For contour detection and speed estimation, the wave does not have to be particularly fast, because both tasks could possibly be performed within milliseconds. In contrast, for region encoding of the SAT a wave needs to propagate rapidly. We argued that fast waves may run through cortical areas (section 7.5).

On the other hand, if the visual system used something like contour-propagation fields, then a wave would not have to be as fast due to the wide-spread representation of the shape. The idea that the brain operates according to some broadcast receiver principle, is certainly unusual too for contemporary computational neuroscience; but given that we have solved some tasks in a relatively simple way with waves, we can only encourage other neuroscientists to consider this viewpoint as well.

The neuronal models we have used in our neuromorphic simulations are merely integrate-and-fire neurons, which are embedded into excitable maps (propagation maps). Their parameter values were sometimes tuned to make the model operate as coincidence detectors that sense when two waves collide (chapter 7). Again, this is not to say that the neuron's function can be reduced to this simple model. The neuron's anatomical and physiological diversity may well be the cause for a variety of distinct wave propagation characteristics.

12.4 Neuromorphic Implementation

We have not presented any analog hardware implementation of our envisioned wave-propagation mechanisms. But they are conceptually simple enough to be implementable with the existing 'silicon ingredients', as for example the circuits presented in chapter 4, which mimic synaptic responses, dendritic propagation and somatic spiking. The wiring substrate, that would enable communication between maps, also already exists: a multichip architecture - the silicon cortex - provides the 'fluent' communication between analog chips (section 4.5). A first step towards a neural hardware realization of our networks would be to build a wave-propagating map as presented in section 6.3, from which one would derive the various variants. It is this analog hardware that allows for a time- and energy-efficient emulation of our wave-propagating maps. The computation of such wave maps in digital computers is and will always be too slow or too energy-consuming. It is therefore the neuromorphic hardware approach that offers the best solution to our envisioned wave networks.

12.5 Future Approach

Short term One short-term goal is to refine some of the map simulations presented in this essay and I am currently in the process of doing so. A next step would be to implement those maps. But even if some of those proposed maps are implemented, it likely requires another round of simulations to ensure that they properly operate in the silicon cortex. This may not be only a minor technical issue, but may also require extensive adjustment and tuning of the dynamics of the respective maps. For example, the SAT architecture (figure 36) or the CPFM system (figure 51), seem straightforward at first, but may bear some intricacies regarding the matching of the map dynamics. Another short-term goal should therefore be the establishment of a soft-ware simulation methodology that guarantees that the envisioned architecture is also transferable into a silicon cortex system.

Long term The challenge of finding the loose basic-level category representations is too vast, that one can give a detailed, meticulous plan on how to proceed. But we have given two broad approaches which may benefit from each other, or may be even converge. One approach was envisioned as the hybrid categorization system in which a neuromorphic front-end extracts contours and encodes regions by the SAT; a computer vision back-end would associate the generated sym-axes (chapter 7). This system can be particularly useful for exploring certain 'high-level' aspects; like the degree of looseness necessary for representing the part-alignment variability and part redundancy. After this looseness has been further characterized, the appropriate networks can be designed. The existent scene recognition approaches (section 2.3.4) may thereby be helpful in determining this looseness - even though they do not exploit the idea of encoding space as we have pursued it here. Thus, we regard the combined employment of computer vision methods and neuromorphic methods as a possibly fruitful approach to explore and develop the future architecture necessary for recognition.

The second approach is the pursuit of a pure neuromorphic system using the CPFM system as a basis (chapters 9 and 10). This approach encodes the region completely as opposed to the SAT. The system learns simple shapes in a single pass and recognizes them even under noisy circumstances. If this network can be extended to basic-level objects, may be with insight from the hybrid approach, then that would be a formidable starting point for the construction of a self-organizing neuromorphic visual system. The most reasonable route would be to extend this system and to make it succeed on line-drawing objects similar to the ones used in chapter 5; then one would further refine the system and make it functioning with gray-scale image input.

Thus, regarding the search for basic-level category representations, the Odysee continues in some sense. But a significant part of it has been completed by constructing networks that encode space. The remainder of the Odysee is now rather a directed one, an exploration towards associating regions somehow.

Terminology

In the vision literature The vision literature (e.g. (Palmer, 1999)) uses a variety of expressions to circumscribe representations and recognition processes, some of which are difficult to define or distinguish from each other. In figure 63 we attempt to organize and clarify those expressions a bit: the left side shows an axis labeled with 'elementary' and 'composite' at the bottom and top respectively. Probably most vision researchers would categorize descriptions and representations along such an axis and would assume that the object- or scene-reconstruction process occurs along that axis somehow, for example into one direction only, or along both directions at the same time.

It is common to use terms like 'features', 'regions', 'parts', 'objects' and 'scenes' which may be ordered according to that axis. The term 'structure' is generally associated with structural description. **Structural description** is the representation format that describes an object as made of well-specified elements. The term structure therefore rather applies to parts, objects and scenes, which are often considered to be made of elementary features. The terms **local and global** are tendentially used by neuroscientists and psychologists. The terms refer to the size of the structure in the visual field and that idea leans very much on the receptive field concept (see also figure 9). The terms **whole and part** are primarily used by psychologists to describe the composition of a structural description, which in some sense represents a nesting of the 'russian puppets', because there is always another part on a more elementary level. The terms **coarse and fine** describe the resolution of the visual image and those terms are not to be confused with the terms global and local (see also (Oliva and Schyns, 1997) for a clarification). On each level of resolution, a separate local/global processing can take place. Both axes can be regarded as partially aligned with the elementary/composite axis. Often, a hierarchical organization is associated with most of the axes and thus the terms **bottom-up** and **top-down** come along with it, expressing the direction of information flow: higher levels of the hierarchy are generally believed to contain more abstract information, lower areas contain (or extract) simpler information.

Neuroscientists tend to think of representation and evolvement as a local-to-global process (chapter 3). Psychologists are influenced by neuroscientific insight, but also by purely computational reflections like the global-to-local concept, which roots in Gestaltists ideas.

In our discourse We use some of the above terms to describe our architectures and components of it, but we do not intend to promote any specific terms, nor do we see any of those evolvement or representation axes rigidly connected to our network. For example in the

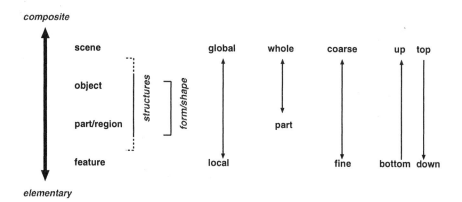

Figure 63: Expressions used in the vision literature to describe representation and reconstruction. See also figure 9.

line-drawing studies with context (section 5.2) made in chapter 5, one can regard a region, like a polygon feature with context analysis, as an elementary feature, but which can extend across large parts of the image, like the Z feature of a chair: hence that feature is of rather global than local nature. Or the CPFM system described in chapter 9.5: the CPF is made of local orientations, but the propagation process does not go along with any of those axes. The following specifies a few terms:

Evolvement   The process of unfolding. I prefer this term to describe the reconstruction or recognition process, because it is less assuming about its exact nature. (The term reconstruction rather implies a systematic, step-wise recovery).

Perceptual Category Representation   Category representations that are loose, flexible and sloppy and that enable quick categorization without detailed analysis. One can regard these representations as the beginning of the evolvement process leading to 'cognition' for example.

Region   The inside and/or outside area of a shape. We refer to the outside also as to the 'silhouette'.

Shape   Simple, two-dimensional form, whose contours can be closed (connected) or open (unconnected).

Space   Synonymous with region, two-dimensional region.

Structure   The usage of this term is commonly restricted to contour description. We relax the term somewhat and include the space (or region), for example like the polygon features used in chapter 5.

Template   Largely fixed representation, which however may contain a small degree of flexibility in order to deal with structural variability.

Visual Exploration   The process of exploring a scene by saccades without a specific motivation.

Visual Search   Commonly used for visual exploration but we understand it as a more goal-oriented search, for example trying to find an object.

References

Adelson, E. and Bergen, J. (1985). Spatiotemporal energy models for the perception of motion. *J. Opt. Soc. Am.*, A2:284–292.

Agin, G. J. and BINFORD, T. (1976). Computer description of curved objects. *IEEE Transactios of Computers*, C-25(4):439–449.

Albright, T. and Stoner, G. (2002). Contextual influences on visual processing. *ANNUAL REVIEW OF NEUROSCIENCE*, 25(339):379–43.

Amit, Y. (2002). *2D object detection and recognition: models, algorithms and networks.* MIT Press, Cambridge, MA.

Baddeley, R., Abbott, L. F., Booth, M. C., Sengpiel, F., Freeman, T., Wakeman, E. A., and Rolls, E. T. (1997). Responses of neurons in primary and inferior temporal visual cortices to natural scenes. *Proc R Soc Lond B Biol Sci*, 264(1389):1775–83.

Barch, D. and Glaser, D. (2002). Slowly moving stimuli induce characteristic periodic activity waves in an excitable membrane model of visual motion processing. *Neurocomputing*, 44:43–50.

Barlow, H. (1953). Summation and inhibition in the frogs retina. *J. Physiol.-London*, 1(119):69–88.

Barlow, H. and Levick, W. (1964). Mechanism of directly selective units in rabbits retina. *J. Physiol. (London)*, 1:477–504.

Barlow, H. B. (1972). Single units and sensation: a neuron doctrine for perceptual psychology? *Perception*, 1:371–394.

Beggs, J. and Plenz, D. (2003). Neuronal avalanches in neocortical circuits. *JOURNAL OF NEUROSCIENCE*, 23(35):11167–11177.

Biederman, I. (1972). Perceiving real-world scenes. *Science*, 177(43):77–80.

Biederman, I. (1987). Recognition-by-components: a theory of human image understanding. *Psychol Rev*, 94(2):115–47.

Biederman, I., Glass, A. L., and r. Stacy EW, J. (1973). Searching for objects in real-world scences. *J Exp Psychol*, 97(1):22–7.

Binford, T. (1971). Visual perception by computer. In *Proceedings of the IEEE Conf. on Systems and Control*, Miami.

Blum, H. (1967). A transformation for extracting new descriptors of shape. In Wathen-Dunn, W., editor, *Models for the Perception of Speech and Visual Form.* MIT Press, Cambridge Mass.

Blum, H. (1973). Biological shape and visual science. *J Theor Biol*, 38(2):205–87.

Blum, H. and Nagel, R. (1978). Shape description using weighted symmetric axis features. *Pattern Recognit.*, 10:167–180.

Boahen, K. (2002). A retinomorphic chip with parallel pathways: Encoding increasing, on, decreasing, and off visual signals. *Analog Integr. Circuits Process.*, 30(2):121–135.

Bradski, G. and Grossberg, S. (1995). Fast-learning viewnet architectures for recognizing three- dimensional objects from multiple two-dimensional views. *Neural Netw.*, 8(7-8):1053–1080.

Brooks, R. (1981). Symbolic reasoning among 3-d models and 2-d images. *Artificial Intelligence*, 17:285–348.

Bruce, V., Green, P., Georgeson, M., and Denbury, J. (2003). *Visual Perception: Physiology, Psychology and Ecology.* Psychology Pr, East Sussex,UK.

Bryant, D. J. and Tversky, B. (1999). Mental representations of perspective and spatial relations from diagrams and models. *J Exp Psychol Learn Mem Cogn*, 25(1):137–56. (eng).

Burbeck, C. and Pizer, S. (1995). Object representation by cores - identifying and representing primitive spatial regions. *Vision Res.*, 35(13):1917–1930.

Burl, M., Leung, T., Weber, M., and Perona, P. (2001). Recognition of visual object classes. In *From Segmentation to Interpretation and Back: Mathematical Methods in Computer Vision.* Springer Verlag.

Canny, J. (1986). A computational approach to edge-detection. *IEEE Transactions on Pattern Analysis and Machine Intelligence*, 8(6):679–698.

Cavanaugh, J., Bair, W., and Movshon, J. (2002). Nature and interaction of signals from the receptive field center and surround in macaque v1 neurons. *JOURNAL OF NEUROPHYSIOLOGY*, 88(5):2530–2546.

Charles, A. (1998). Intercellular calcium waves in glia. *Glia*, 24(1):39–49.

Clowes, M. B. (1971). On seeing things. *Artificial Intelligence*, 2(1):79–116.

Dayan, P. and Abbott, L. (2001). *Theoretical Neuroscience: Computational and Mathematical Modeling of Neural Systems*. MIT Press, Cambridge, Mass.

De Valois, R. L. and De Valois, K. K. (1988). *Spatial Vision*. Oxford University Press, New York.

deCharms, R. C. and Zador, A. (2000). Neural representation and the cortical code. *Annual Review of Neuroscience*, 23:613–647.

Deiss, S., Douglas, R., and Whatley, A. (1999). A pulse-coded communications infrastructure for neuromorphic systems. In Maass, W. and Bishop, C. M., editors, *Pulsed Neural Networks*, chapter 6, pages 157–178. The MIT Press.

Deutsch, J. (1962). A system for shape recognition. *Psychological Review*, 69(6):492–500.

Dill, A. and Levine, M. (1987). Multiple resolution skeletons. *IEEE Trans. Pattern Anal. Mach. Intell.*, 9(4):495–504.

Douglas, R., Mahowald, M., and Mead, C. (1995). Neuromorphic analog VLSI. *Ann. Rev. Neurosci.*, 18:255–281.

Dowling, J. (1987). *The Retina: An Approachable Part of the Brain*. Harvard University Press, Cambridge, MA.

Draper, B., Hanson, A., and Riseman, E. (1996). Knowledge-directed vision: Control, learning, and integration. *PROCEEDINGS OF THE IEEE*, 84(11):1625–1637.

Duda, R. O. and Hart, P. E. (1973). *Pattern Classification and Scene Analysis*. John Wiley and Sons Inc.

Edelman, S. (1999). *Representation and Recognition in Vision*. MIT Press, Cambridge, MA.

Elias, J. G. (1993). Artificial dendritic trees. *Neural Computation*, 5:648–664.

Ermentrout, G. and Kleinfeld, D. (2001). Traveling electrical waves in cortex: insights from phase dynamics and speculation on a computational role. *Neuron*, 29(1):33–44.

Farah, M. J. (1990). *Visual Agnosia*. MIT Press, Cambridge, Massachusetts.

Farah, M. J. (2000). *The Cognitive Neuroscience of Vision*. Blackwell Pub, Malden, Mass.

Felleman, D. and Van Essen, D. (1991). Distributed hierarchical processing in the primate cerebral cortex. *Cerebral Cortex*, 1:123–200.

Francis, G., Grossberg, S., and Mingolla, E. (1994). Cortical dynamics of feature binding and reset - control of visual persistence. *Vision Res.*, 34(8):1089–1104.

Freedman, D., Riesenhuber, M., Poggio, T., and Miller, E. (2002). Visual categorization and the primate prefrontal cortex: Neurophysiology and behavior. *JOURNAL OF NEUROPHYSIOLOGY*, 88(2):929–941.

Fukushima, K. (1988). Neocognitron - a hierarchical neural network capable of visual- pattern recognition. *Neural Netw.*, 1(2):119–130.

Gabor, D. (1946). Theory of communication. *J. Inst. Elect. Eng. (Lond.)*, 93:429–457.

Galarreta, M. and Hestrin, S. (1999). A network of fast-spiking cells in the neocortex connected by electrical synapses. *Nature*, 402(6757):72–5.

Gallant, J. L., Connor, C. E., Rakshit, S., Lewis, J. W., and Essen, D. C. V. (1996). Neural responses to polar, hyperbolic, and cartesian gratings in area v4 of the macaque monkey. *J Neurophysiol*, 76(4):2718–39.

GERRISSEN, J. (1982). Theory and model of the human global analysis of visual structure. *IEEE TRANSACTIONS ON SYSTEMS MAN AND CYBERNETICS*, 12(6):805–817.

GERRISSEN, J. (1984). Theory and model of the human global analysis of visual structure .2. the space-time and visual value segment. *IEEE TRANSACTIONS ON SYSTEMS MAN AND CYBERNETICS*, 14(6):847–862.

Gerstner, W. and Kistler, M. (2002). *Spiking Neuron Models*. Cambridge University Press.

Gibson, J. J. (1950). *The Perception of the Visual World*. Houghton Mifflin, Boston.

Gibson, J. R., Beierlein, M., and Connors, B. W. (1999). Two networks of electrically coupled inhibitory neurons in neocortex. *Nature*, 402(6757):75–9.

Glaser, D. and Barch, D. (1999). Motion detection and characterization by an excitable membrane: The 'bow wave' model. *Neurocomputing*, 26-7(jun):137–146.

Gregory, R. (1997). Knowledge in perception and illusion. *Philos. Trans. R. Soc. Lond. Ser. B-Biol. Sci.*, 352(1358):1121–1127.

Grimson, W. E. L. (1990). *Object recognition by computer: the role of geometric constraints*. MIT Press, Cambridge, Mass.

Grinvald, A., Lieke, E., Frostig, R., and Hildesheim, R. (1994). Cortical point-spread function and long-range lateral interactions revealed by real-time optical imaging of macaque monkey primary visual-cortex. *J. Neurosci.*, 14(5):2545–2568.

Gross, C., Rocha-Miranda, C., and Bender, D. (1972). Visual properties of neurons in inferotemporal cortex of the macaque. *J. Neurophysiol.*, 35:96–111.

Guzman, A. (1969). Decomposition of a visual scene into three-dimensional bodies. In Grasselli, editor, *Automatic Interpretation and Classification of Images*, chapter 12. Academic Press, New York.

Guzman, A. (1971). Analysis of curved line drawings using context and global information. In Meltzer, M. and Michie, D., editors, *Machine Intelligence 6*, chapter 20, pages 325–375. Edingburgh University Press, Edingburgh, Scotland.

Häfliger, P. and Rasche, C. (1999). floating gate analog memory for parameter and variable storage in a learning silicon neuron. In *IEEE International Symposium on Circuits and Systems*, Orlando.

Haykin, S. (1994). *Neural Networks: A Comprehensive Foundation*. Prentice Hall.

Hegde, J. and Essen, D. C. V. (2000). Selectivity for complex shapes in primate visual area v2. *J Neurosci (Online)*, 20(5):RC61.

Henderson, J. M. and Hollingworth, A. (1999). High-level scene perception. *Annu Rev Psychol*, 50:243–71.

Hess, R. and Field, D. (1999). Integration of contours: new insights. *TRENDS COGN. SCI.*, 3(12):480–486.

Hodgkin, A. and Huxley, A. (1952). A quantitative description of membrane current and its application to conduction and excitation in nerve. *Journal of Physiology*, 117:500–544.

Hopfield, J. (1995). Pattern recognition computation using action potential timing for stimulus representation. *Nature*, 376:33–36.

Hubel, D. and Wiesel, T. (1962). Receptive fields, binocular interaction and functional architecture in the cat's visual cortex. *J. Physiol.*, 160:106–154.

Hubel, D. and Wiesel, T. (1968). Receptive fields and functional architecture of monkey striate cortex. *J. Physiol. (London)*, 195:215–243.

Huffman, D. (1971). Impossible objects as nonsense sentences. In Meltzer, M. and Michie, D., editors, *Machine Intelligence 6*, chapter 19, pages 295–323. Edingburgh University Press, Edingburgh, Scotland.

Hughes, J. (1995). The phenomenon of traveling waves - a review. *Clin. Electroencephalogr.*, 26(1):1–6.

Humphreys, G. and Riddoch, M. (1987a). The fractionation of visual agnosia. In Humphreys, G. and Riddoch, M., editors, *Visual Object Processing*, chapter 10, pages 281–306. Lawrence Erlbaum Associates.

Humphreys, G. and Riddoch, M. (1987b). *To See But Not to See: A Case Study of Visual Agnosia*. Lawrence Erlbaum Associates, London.

Indiveri, G. (2001). A neuromorphic vlsi device for implementing 2-d selective attention systems. *IEEE TRANSACTIONS ON NEURAL NETWORKS*, 12(6):1455–1463.

Itti, L. and Koch, C. (2000). A saliency-based search mechanism for overt and covert shifts of visual attention. *Vision Res*, 40(10-12):1489–506.

Jacobs, A. and Werblin, F. (1998). Spatiotemporal patterns at the retinal output. *J. Neurophysiol.*, 80(1):447–451.

Johnston, D., Magee, J. C., Colbert, C. M., and Cristie, B. R. (1996). Active properties of neuronal dendrites. *Annu Rev Neurosci*, 19:165–86.

Johnston, D. and Wu, S. (1995). *Foundations of Cellular Neurophysiology*. MIT, Cambridge, Massachusetts.

Kegl, B. and Krzyak, A. (2002). Piecewise linear skeletonization using principal curves. *IEEE Trans. Pattern Anal. Mach. Intell.*, 24(1):59–74.

Koch, C. (1999). *Computational Biophysics of Neurons*. MIT, Cambridge:Mass.

Koch, C. (2004). *The Quest for Consciousness: A Neurobiological Approach*. Roberts and Co.

Koch, C., Poggio, T., and Torre, V. (1983). Nonlinear interactions in a dendritic tree: localization, timing, and role in information processing. *Proc Natl Acad Sci U S A*, 80(9):2799–802.

Koch, C. and Ullman, S. (1985). Shifts in selective visual attention: towards the underlying neural circuitry. *Hum Neurobiol*, 4(4):219–27.

Koenderink, J. J. and van Doorn, A. J. (1986). Depth and shape from differential perspective in the presence of bending deformations. *J Opt Soc Am A*, 3(2):242–9. (eng).

Koenig, P., Engel, A., and Singer, W. (1996). Integrator or coincidence detector? the role of the cortical neuron revis ited. *TINS*, 19:130–137.

Koffka, K. (1935). *Principles of Gestalt Psychology*. Harcourt, Brace, New York.

Kovacs, I. (1996). Gestalten of today: Early processing of visual contours and surfaces. *Behav. Brain Res.*, 82(1):1–11.

Kovacs, I. and Julesz, B. (1994). Perceptual sensitivity maps within globally defined visual shapes. *Nature*, 370(6491):644–6.

Kramer, J., Sarpeshkar, R., and Koch, C. (1995). An analog VLSI velocity sensor. In *Int. Symposium on Circuits and Systems*, pages 413–416. IEEE.

Kreiman, G., Koch, C., and Fried, I. (2000). Category-specific visual responses of single neurons in the human medial temporal lobe [see comments]. *Nat Neurosci*, 3(9):946–53. Comment in: Nat Neurosci 2000 Sep;3(9):855-6.

Kuffler, S. (1953). Discharge patterns and functional organization of mammalian retina. *J. Neurophysiology*, 16:37–68.

Lee, H. and Fu, K. (1981). The glgs image representation and its application to preliminary segmentation and pre-attentive visual search. In *Proceedings of the IEEE Comput. Soc. Conf. Patterns Recognition and Image Processing*, pages 256–261.

Lee, H. and Fu, K. (1983). Generating object descriptions for model retrieval. *IEEE Trans. Pattern Anal. Mach. Intell.*, 5(5):462–471.

Lee, T., Mumford, D., Romero, R., and Lamme, V. (1998). The role of the primary visual cortex in higher level vision. *Vision Res.*, 38(15-16):2429–2454.

Leymarie, F. and Levine, M. (1992). Simulating the grassfire transform using an active contour model. *IEEE Trans. Pattern Anal. Mach. Intell.*, 14(1):56–75.

Liu, S., Kramer, J., Indiveri, G., Delbruck, T., Burg, T., and Douglas, R. (2001). Orientation-selective avlsi spiking neurons. *NEURAL NETWORKS*, 14(6-7):629–643.

Livingstone, M. (1998). Mechanisms of direction selectivity in macaque v1. *NEURON*, 20(3):509–526.

Lowe, D. (1987). 3-dimensional object recognition from single two- dimensional images. *Artificial Intelligence*, 31(3):355–395.

Luck, S. J., Chelazzi, L., Hillyard, S. A., and Desimone, R. (1997). Neural mechanisms of spatial selective attention in areas v1, v2, and v4 of macaque visual cortex. *J Neurophysiol*, 77(1):24–42.

Mahowald, M. and Mead, C. (1991). Silicon retina. *Scientific American*, 264(5):76–82.

Malik, J. and Perona, P. (1990). Preattentive texture-discrimination with early vision mechanisms. *J. Opt. Soc. Am. A-Opt. Image Sci. Vis.*, 7(5):923–932.

Marr, D. (1982). *Vision*. W. H. Freeman, New York.

Marr, D. and Nishihara, H. (1978). Representation and recognition of spatial-organization of 3- dimensional shapes. *Proc. R. Soc. Lond. Ser. B-Biol. Sci.*, (1140):269–294.

McCulloch, W. (1965). *Embodiments of Mind*. MIT Press, Cambridge, Massachusetts.

Mead, C. A. (1989). *Analog VLSI and Neural Systems*. Addison-Wesley, Reading, Massachusetts.

Mel, B. W. (1993). Synaptic integration in an excitable dendritic tree. *J Neurophysiol*, 70(3):1086–101.

Minsky, M. (1975). A framework for representing knowledge. In Winston, P., editor, *The Psychology of Computer Vision*, pages 211–277. McGraw-Hill, New York.

Moran, J. and Desimone, R. (1985). Selective attention gates visual processing in the extrastriate cortex. *Science*, 229(4715):782–4.

Nakayama, K. (1985). Biological image motion processing - a review. *VISION RESEARCH*, 25(5):625–660.

Navon, D. (1977). Forest before trees: the precedence of global features in visual perception. *Cognitive Psychology*, 9:353–383.

Neisser, U. (1967). *Cognitive Psychology*. Appleton-Century-Crofts, New York.

Nowlan, S. J. and Sejnowski, T. J. (1995). A selection model for motion processing in area mt of primates. *J Neurosci*, 15(2):1195–214.

Ogniewicz, R. and Kubler, O. (1995). Voronoi tessellation of points with integer coordinates: Time- efficient implementation and online edge-list generation. *Pattern Recognit.*, 28(12):1839–1844.

Oliva, A. and Schyns, P. G. (1997). Coarse blobs or fine edges? evidence that information diagnosticity changes the perception of complex visual stimuli. *Cognit Psychol*, 34(1):72–107.

O'Regan, J. K. (1992). Solving the "real" mysteries of visual perception: the world as an outside memory. *Can J Psychol*, 46(3):461–88.

O'Regan, J. K., Rensink, R. A., and Clark, J. J. (1999). Change-blindness as a result of 'mudsplashes' [letter]. *Nature*, 398(6722):34.

Palmer, S. E. (1999). *Vision Science: Photons to Phenomenology*. MIT Press, Cambridge, Massachusetts.

Palmer, S. E., Rosch, E., and Chase, P. (1981). Canonical perspective and the perception of objects. In Long, J. and Baddeley, A., editors, *Attention and performance IX*, pages 135–151. Erlbaum, Hillsdale, NJ.

Parasuraman, R. (1998). *The attentive brain*. MIT Press, Cambridge, Massachusetts.

Pasupathy, A. and Connor, C. (1999). Responses to contour features in macaque area v4. *JOURNAL OF NEUROPHYSIOLOGY*, 82(5):2490–2502.

Pentland, A. (1986). Perceptual organization and the representation of natural form. *Artificial Intelligence*, 28:293–331.

Perrone, J. and Thiele, A. (2001). Speed skills: measuring the visual speed analyzing properties of primate mt neurons. *NATURE NEUROSCIENCE*, 4(5):526–532.

Pizer, S., Oliver, W., and Bloomberg, S. (1987). Hierarchical shape-description via the multiresolution symmetrical axis transform. *IEEE Trans. Pattern Anal. Mach. Intell.*, 9(4):505–511.

Potter, M. C. (1975). Meaning in visual search. *Science*, 187(4180):965–6.

Potter, M. C. (1976). Short-term conceptual memory for pictures. *J Exp Psychol [Hum Learn]*, 2(5):509–22.

Prechtl, J., Cohen, L., Pesaran, B., Mitra, P., and Kleinfeld, D. (1997). Visual stimuli induce waves of electrical activity in turtle cortex. *Proc. Natl. Acad. Sci. U. S. A.*, 94(14):7621–7626.

Priebe, N., Cassanello, C., and Lisberger, S. (2003). The neural representation of speed in macaque area mt/v5. *JOURNAL OF NEUROSCIENCE*, 23(13):5650–5661.

Psotka, J. (1978). Perceptual processes that may create stick figures and balance. *J. Exp. Psychol.-Hum. Percept. Perform.*, 4(1):101–111.

Rall, W. (1964). Theoretical significance of dendritic trees for neuronal input-output relations. In Reiss, R., editor, *Neural theory and modelling*, pages 73–97. Stanford:Stanford University Press.

Rasche, C. (1999). *Analog VLSI Circuits for Emulating Computational Features of Pyramidal Cells*. PhD thesis, No. 13268, Eidgenössische Technische Hochschule Zürich, Switzerland.

Rasche, C. (2002a). Structural description with classified polygon and surface features and their groupings. *search http://www.springeronline.com/ for book, see surfpoly.pdf.*

Rasche, C. (2002b). Structural description with feature context. *search http://www.springeronline.com/ for book, see context.pdf.*

Rasche, C. (2004). Signaling contours by neuromorphic wave propagation. *Biological Cybernetics*, 90(4):272–279.

Rasche, C. (2005a). Excitable maps for visual processing. *under review.*

Rasche, C. (2005b). A neuromorphic simulation of the symmetric axis transform. *under review.*

Rasche, C. (2005c). Speed estimation with propagation maps. *under review.*

Rasche, C. (2005d). Visual shape recognition with contour propagation. *under review.*

Rasche, C. and Douglas, R. (1999). Silicon synaptic conductances. *Journal of Computational Neuroscience*, 7:33–39.

Rasche, C. and Douglas, R. (2001). Forward- and backpropagation in a silicon dendrite. *IEEE Transactions on Neural Networks*, 12(2):386–393.

Rasche, C. and Hahnloser, R. (2001). Silicon synaptic depression. *Biological Cybernetics*, 84(1):57–62.

Rasche, C. and Koch, C. (2002). Recognizing the gist of a visual scene: possible perceptual and neural mechanisms. *Neurocomputing*, 44-46:979–984.

Reichardt, W. (1961). Autocorrelation, a principle for the evaluation of sensory information by the central nervous system. In Rosenblith, W., editor, *Sensory Communication*, pages 303–317. Wiley, New York.

Rensink, R. A. (2000). Seeing, sensing, and scrutinizing. *Vision Res*, 40(10-12):1469–1487. (ENG).

Riesenhuber, M. and Poggio, T. (1999). Hierarchical models of object recognition in cortex. *Nat Neurosci*, 2(11):1019–25.

Robert, L. G. (1965). Machine perception of three-dimensional solids. In Tippet, T. e. a., editor, *Optical and Electro optical Information Processing*. MIT Press.

Rolls, E. and Deco, G. (2002). *Computational neuroscience of vision*. Oxford University Press, New York.

Rosch, E., Mervis, C., Gray, W., and Boyes-Braem, P. (1976). Basic objects in natural categories. *Cognitive Psychology*, 8:382–439.

Rullen, R. V., Gautrais, J., Delorme, A., and Thorpe, S. (1998). Face processing using one spike per neurone. *Biosystems*, 48(1-3):229–39.

Schendan, H. E., Ganis, G., and Kutas, M. (1998). Neurophysiological evidence for visual perceptual categorization of words and faces within 150 ms. *Psychophysiology*, 35(3):240–51.

Shepard, G. (1998). *The Synaptic Organization of the Brain*. Oxford University Press, New York, NY, 4 edition.

Shevelev, I. and Tsicalov, E. (1997). Fast thermal waves spreading over the cerebral cortex. *Neuroscience*, 76(2):531–540.

Singer, W., Artola, A., König, P., Kreiter, A., Lowel, S., and T.B., S. (1993). Neuronal representations and temporal codes. In Poggio, T. and Glaser, D., editors, *Exploring Brain Functions: Models in Neuroscience*, chapter 13, pages 179–194. John Wiley, Chichester: UK.

Standing, L., Conezio, J., and Haber, R. (1970). Perception and memory for pictures - single-trial learning of 2500 visual stimuli. *PSYCHONOMIC SCIENCE*, 19(2):73–74.

Sugita, Y. (1999). Grouping of image fragments in primary visual cortex. *Nature*, 401(6750):269–72.

Svoboda, K., Denk, W., Kleinfeld, D., and Tank, D. W. (1997). In vivo dendritic calcium dynamics in neocortical pyramidal neurons. *Nature*, 385(6612):161–5.

Tanaka, K. (1996). Inferotemporal cortex and object vision. *Annual Review of Neuroscience*, 19:109–139.

Tanaka, K., Sugita, Y., Moriya, M., and Saito, H. (1993). Analysis of object motion in the ventral part of the medial superior temporal area of the macaque visual cortex. *J Neurophysiol*, 69(1):128–42.

Tanner, J. E. and Mead, C. (1986). An integrated analog optical motion sensor. In *et al*, S.-Y. K., editor, *VLSI Signal Processing, II*, pages 59–76, New York. IEEE Press.

Tarr, M. J. and Bulthoff, H. H. (1998). Image-based object recognition in man, monkey and machine. *Cognition*, 67(1-2):1–20. (eng).

Thorpe, S., Fize, D., and Marlot, C. (1996). Speed of processing in the human visual system. *Nature*, 381:520–522.

Thorpe, S. J. (1990). Spike arrival times: a highly efficient coding scheme for neural networks. In Eckmiller, R., Hartmann, G., and Hauske, G., editors, *Parallel Processing in Neural Systems and Computers*, pages 91–94. Elsevier Science Publishers.

Trappenberg, T. (2002). *Fundamentals of Computational Neuroscience*. Oxford University Press, New York, NY, 1 edition.

Treisman, A. (1988). Features and objects: the fourteenth bartlett memorial lecture. *Q J Exp Psychol [A]*, 40(2):201–37.

Tuckwell, H. C. (2000). Cortical potential distributions and information processing [in process citation]. *Neural Comput*, 12(12):2777–95.

ULLMAN, S. (1990). 3-dimensional object recognition. *COLD SPRING HARBOR SYMPOSIA ON QUANTITATIVE BIOLOGY*, 55:889–898.

Ullman, S. and Sali, E. (2000). Object classification using a fragment-based representation. *BIOLOGICALLY MOTIVATED COMPUTER VISION, PROCEEDING*, 1811:73–87.

Vansanten, J. and Sperling, G. (1984). Temporal covariance model of human motion perception. *JOURNAL OF THE OPTICAL SOCIETY OF AMERICA A-OPTICS IMAGE SCIENCE AND VISION*, 1(5):451–473.

Vinje, W. E. and Gallant, J. L. (2000). Sparse coding and decorrelation in primary visual cortex during natural vision. *Science*, 287(5456):1273–6.

Waltz, D. (1975). Understanding line drawings of scenes with shadows. In Winston, P., editor, *The Psychology of Computer Vision*. McGraw-Hill, New York.

Watson, A. and Ahumada, A. (1985). Model of human visual-motion sensing. *JOURNAL OF THE OPTICAL SOCIETY OF AMERICA A-OPTICS IMAGE SCIENCE AND VISION*, 2(2):322–342.

Wilson, H., Blake, R., and Lee, S. (2001). Dynamics of travelling waves in visual perception. *Nature*, 412(6850):907–910.

Yarbus, A. (1967). *Eye movements and vision*. Plenum Press, New York.

Index

Keywords

A

agnosia	29,9
analog circuits	37
apperture problem	84
artificial intelligence	3,10
attention	4,21,21
attenional shifts	23

B

basic-level category	ii,3,94
biophysical plausibility	74,83,92,107
binding	20,51
bottom-up	19,107
broadcast receiver	26,93

C

canonical (view)	7,45
category representation	7
change blindness	115
channel (ionic)	33
channel (visual)	27
channel (transistor)	37
classical receptive field	21,
classifier	119
cognition	3
coincidence detector, neuron	32,68
compartmental modeling	39,58,78
computer vision	10,45,60,71,87
context	49
contour extraction	55,64,121
contour grouping	73,121
contour propagation	27
contour propagation field	97
cortical potential distributions	26

D

delay-and-compare principle	78
dendrite	32
dendritic compartments	33
dendro-dendritic	83
differentiation	108
direction selectivity	78,98
distributed coding and hierarchy	20
dynamic representation	91

E

end-stopped cells	88
engineering goal	5
excitatory postsynaptic potential	32
event-related potential	24
evolvement	2,9

F

fabrication noise	42
feature integration	18
field potential	26
filter	22
Fourier transform	22,67
frame	3,13,14,24,93

G

ganglion cells	17,55
geographical structure, map	55,64
Gestalt	27,125

gist   1,25,111
glia   57,84
global-to-local   25,51,125
gray-scale image   55

H
hiearchy   17,125
heterarchy (cortical)   20
hybrid categorization system   76,122

I
illusion   2,113,79
inferior temporal cortex (IT)   18
integrate-and-fire neuron   34

L
latency code   24
learning   107
L feature   45,71,77,87
line drawing   13,45,67,31
local-to-global   17,51,106,125
luminance landscape   55

M
magnetic field   26
memory capacity   5
morse code   25
motion detection   77,87
multi-chip architecture   40

N
natural stimulus   see stimulus natural
neural code   25, see also latency code,
                                                rate code and timing code
noise   42,73,105
non-canonical   7

O
orientation columns   18,40,68
optical flow field   97

P
paradox   5
part-alignment variability   7
part-shape variability   7
part redundancy   7
perception   3
perceptual category representations   3,125
polygon feature   45
polyhedra   13
position invariance   89
propagating wave   see wave propagation
primal sketch   11
pyramidal architecture   89,see also speed pyramid

R
rapid-serial-visual-presentation   1
rate code   18,25,55
RC circuit   31
reconstruction   9
region   125
resistor-capacitor circuit   see RC circuit just above
receptive field   17,69,106
region encoding   27,68,88
response time   105
retina   17,55
rotation invariance   91

S

saccades   9,23,92,113
saliency map   113
scale invariance   see size invariance
scene exploration   113
scene recognition   13,111
self-interacting shape   27
shape   125
shape map   98
silhouette feature   49
silicon neuron   40
silicon cortex   40
size invariance   89
spatial frequency   22
speed estimation   79
speed map   80
speed pyramid   80
speed tuning curves   82
stimuli natural   21
structural description   125
structural variability   7
structure   125
subordinate category   3
surface feature   49
sym-ax   see symmetric-axis transform
symmetric-axis transform (SAT)   27,67,105
sym-point   67
synaptic circuit   38
synaptic response   32,79

T
template   13,92
template architecture   92
texture   119
texture segragation   23
timing code   21,25
top-down   3,20,107
traveling wave   same as propagating wave
trajectory   67,77
transistor   37
translation invariance   see position invariance
TV commercial   1

V
very-large-scale integrated circuits   37
viewpoint independence   7
visual exploration   1,113
visual search   1,9,107,113

W
wave propagation   27
wavelet   23,93

Abbreviations

2D, 3D   two-, three-dimensional
aVLSI   analog very-large-scale integrated
CPF   contour propagation field
CPFM   contour propagation field matching
DC(s)   direction columns
EPSP   excitatory postsynaptic potential
ERP   event-related potential
I&F   integrate-and-fire
OC(s)   orientation columns
PM   propagation map
RC   resistor-capacitor
RF   receptive field
SAT   symmetric-axis transform

SAM	symmetric-axis map
SM	shape map
V1	primary visual cortex
V2	higher cortical area
V4	higher cortical area
V5	see MT
IT	inferior temporal cortex
MT	medial temporal cortex
VLSI	very-large-scale-integrated (circuit)